THE FINGER BOOK

The Finger Book

Sex, Behaviour and Disease
Revealed in the Fingers

JOHN T. MANNING

faber and faber

First published in 2008
by Faber and Faber Limited
3 Queen Square London WC1N 3AU

Typeset by RefineCatch Limited, Bungay, Suffolk
Printed in England by Mackays of Chatham, plc

A CIP record for this book
is available from the British Library

ISBN 978–0–571–21539–3

Contents

Preface

What could fingers and sex possibly have in common? The ideas I put forward here represent the end of a personal journey of enquiry – a journey taken without much of a map, devoid of well-defined paths and often without a clearly defined sense of my destination. With this in mind I would like briefly to describe the framework of questions that have driven me along this particular path, and then outline how I have stumbled through this journey led by my list of questions.

In this book I take a tiny sex difference in the fingers – men's tendency to have longer ring fingers relative to their index fingers than women – and use this 'finger ratio' to explain many puzzles about human behaviour and predisposition to disease. The key to the power of this ratio is the point in development at which the sex difference arises, for it is found in very young children and is not accentuated at puberty. Many questions flow from these simple observations. Is the ratio a reflection of hormones such as testosterone and oestrogen in the womb? If so, could it point the way to how early sex hormones lie at the heart of sex differences in the foetus, in the child and in the adult? Many important diseases are different in their incidence and progression in men and women. Is this fixed early, under the influence of sex hormones? Could the finger ratio point to how and why these differences arise? Are differences in behaviour between men and women solely the result of social pressures or are they influenced by sex hormones in the womb? Then there is sexual orientation. Is homosexuality the result of a certain family environment, is it a lifestyle decision, or is its root in our foetal hormonal bath? Then there are the creative aspects of our life. What indelible mark has the mysterious process of hormonal 'foetal programming' left on our culture? Suppose, for example, that music is merely a form of male sexual display. Can the finger ratio tell us what musicians are actually displaying to women? How about male-dominated sports? Are they highly public tests of our foetal development? Finally, what of the great sweep of human evolution? Can this in part be explained by changes in the relative amounts of sex hormones in the womb?

My investigations of the finger ratio have led me to conclude that the answers to these questions must include a consideration of the early sex hormones – testosterone and oestrogen – and their role in the development of the foetus. Some of my arguments are controversial. Some have been called premature because they have not yet been formally tested. However, I set this framework out for you in its entirety for two reasons. The first is that I simply cannot resist the flow of ideas that spring from established fact, circumstantial evidence and intuition. Second, I would like to invite you to inspect this edifice of facts, ideas and enthusiasm and form your own opinions. I will guide you in this, and give warnings when I am describing associations that may be exciting but are as yet unverified. Eventually, I hope my colleagues and I will pick over the model I present, testing each hypothesis and pursuing each theory; for now, this is the best interpretation of the available facts I can offer.

I would like to begin by explaining how I came to the subject. My first love was of animals, and at London University in the late 1960s I delighted in the findings of the great Victorian naturalists and taxonomists such as Richard Owen, Charles Darwin, Alfred Russel Wallace and Thomas Henry Huxley. The work of these and others has led us to an understanding of how animals have evolved and how we can detect the evidence of such evolution from the study of development and anatomy. At the heart of their success in this endeavour is the comparative method. Consider for a moment the amazing variety of form and function in the animal kingdom – how do we make sense of so much diversity? Simply using words such as the Mollusca and the Vertebrata implies a comparative approach. It acknowledges that those before us have grouped animals on the basis of their similarities and differences into phyla, each of which owes its existence to a common ancestral population that has left its mark on all the progeny species. This can be demonstrated by comparing, say, a fish, a frog, a lizard, a bird and a mammal. All have a backbone and are descended from a common ancestral species that also possessed a backbone. The frog, lizard, bird and mammal possess four limbs; these tetrapods are descended from a four-limbed species whose limbs ended with five toes. The members of one group, the true mammals, have internal fertilisation and foetal development within the womb, which they owe to an ancestral species which lived perhaps 70 million years ago. Humans fit readily into this framework, belonging to the mammalian group that is 'first in importance', i.e. the primates. Like all other primates, our bodies show evidence of an ancestral tree-dwelling existence. The comparative method enables us to take this further: we are large for a primate, without a tail, and do not live habitually in trees, and are therefore a species of great ape. Our closest living relatives are the two species of chimpanzee (the bonobo and common chimpanzee) and the gorilla. Today, all three apes are found in Africa, mainly west of the Great Rift Valley. Fossil evidence indicates early humans to have originated to the east of the Rift Valley away from direct competition from their nearest relatives. Thus the comparative method reveals that humans are best considered within the framework of their tetrapod, mammalian and primate origins. This is important when we come to consider the evolution of the finger ratio.

If my first love was animals, my second was the evolution of sex. After my B.Sc. I followed a Masters course in evolution at Liverpool University, whose Life Sciences departments were illuminated by the presence of Arthur Cain, Phillip Sheppard and Tony Bradshaw. Of the three it was Arthur Cain, Professor of Zoology, who had the greatest impact on me. Cain was enthusiastic about evolution and believed implicitly in the power of natural selection as its main driving force. In his lectures the great figures of nineteenth-century biology came to life as he described the passionate clash between scripture and evolution, divine creation and natural selection, 'God's will' and blind chance.

In this environment I developed a keen interest in the evolution of sex, which in turn would lead me to link fingers and sex. Reproduction in animals, particularly complex animals, involves a male and a female, each of whom produces sex cells. Each offspring results from the fusion of two sex cells. Both males and females have two sets of genes: these are replicated, shuffled and then copies of one set of genes are placed in each sex cell. The fusion of two sex cells results in a new individual with two sets of genes. In my eyes the Holy Grail of biology was to be found in the answers to two questions regarding sex. First, why shuffle one's genes before placing them in sex cells? Second, what is the point of males and females? I have been pondering these questions for some thirty-five years. Let us concentrate on the second question, for the answer will lead us to our goal of sex and fingers.

The essence of being a male is the production of sex cells called sperm. Sperm cost the body little to produce, and are tiny and very numerous. This means that males can potentially create huge numbers of offspring. I say potentially because in their struggle to reproduce, males must gain access to the other type of sex cell, the egg. Females produce eggs and guard them fiercely because they are costly to make, large and not numerous. This applies to females of all species, including women. In these simple biological facts lie explanations for both the attraction and the periodic conflict between men and women. Both sexes strive to reproduce in order to transmit copies of their genes to the next generation. In this they must cooperate. However, men may produce many offspring because sperm cost little to make. In order to do this they must have the cooperation of many women, since for women a baby represents an expensive egg, nine months of gestation and metabolically expensive suckling. This limits their reproductive output, forcing them into a strategy that emphasises quality rather than quantity of offspring. Thus men compete against men for women, while women compete against women for the best male genes on offer and for the resources that will enable them to invest in their babies.

This, then, is sexual selection theory. By 1972 it seemed apparent to me that in humans sex differences and sexual behaviour could be explained within this framework. It was then that I started the long journey towards a sex and fingers link. Continuing my studies at Liverpool on a part-time basis, I became convinced by my work on crustaceans, insects, birds, monkeys and apes that males were indeed driven by their biology to compete for fertile females and in turn the imperatives for females were good male genes and resources. By this time I had my Ph.D. and a permanent appointment at Liverpool University, and in 1990 I started my work on sex and symmetry.

For eight years I worked on and enthused about animal symmetry as a simple physical cue for 'good genes'. Humans and other animals tend to be symmetrical in terms of their external characteristics. Thus our right ear is similar in size and shape to our left ear, our right wrist is similar to our left wrist and our right ankle is similar to our left ankle. However, when these and other paired traits are measured we often find small right–left differences of a few millimetres. Women may be able to perceive such small deviations from perfect symmetry, which possibly reflect an underlying developmental instability caused by inefficient or damaged genes. They might then reject the advances of asymmetrical men in favour of those with symmetry and good genes. After eight years of measuring tiny body asymmetries I concluded that the theory was essentially correct for humans. There is now a considerable amount of data supporting the notion that symmetry is related to efficiency of function. Comparisons between asymmetrical and symmetrical men show that the latter have higher IQ, run faster, are less likely to withdraw from fights with other males and produce more and faster-swimming sperm. Women appear to be able to perceive symmetry in men with remarkable precision, rating symmetrical male faces and bodies as attractive, reporting more orgasms with symmetrical partners and even preferring the body odour of symmetrical men. All this relates to choice of sexual partner in adults, but my interests shifted towards the symmetry of children and the importance of early development. At this time

I was involved in the establishment of the Jamaican Symmetry Project.

I spent much of January 1998 in a well-run primary school in the undeveloped south of Jamaica. The result of an initiative by Robert Trivers of Rutgers University, the Jamaican Symmetry Project focused on asymmetries in children. We were concerned initially with detailed measurement of some three hundred children aged from five to eleven years. We worked hard measuring ears, fingers, wrists, elbows, knees, ankles and feet. When this happiest of times for me came to an end, I returned to Liverpool to find that my student Tina Wilson had provided considerable data on asymmetry in local children. From these data it was apparent that at two years of age boys and girls are very asymmetrical, but with further growth asymmetries are reduced until puberty, at which point there is another burst of asymmetry which reduces further, until at eighteen years of age we are at our most symmetrical and presumably most attractive. My feeling was that the explanation for this pattern might be sex hormones, and in particular testosterone and oestrogen.

Puberty, with its welter of sex hormones and major changes in body shape, is well understood. But what simple trait would give me a measure of sex hormones experienced in the foetus? The idea of the second-to-fourth-finger ratio came to me during an asymmetry project on infertility carried out at Liverpool Children's Hospital in collaboration with fertility expert Iwan Lewis-Jones. The asymmetries of ears, wrists and fingers in men and their partners were measured. It was found that the most asymmetrical patients produced fewer and less active sperm than the symmetrical men, although the associations were quite weak. At the time I was reading of new insights regarding the understanding of the genetic control of finger development. Intriguingly, it turned out that the genes that controlled finger development also controlled the formation of the reproductive system. Looking at my data on fingers and sperm, I noted that men's fingers were on average longer than women's: this was just a function of body size. Closer inspection showed that men had longer ring fingers relative to their index fingers than women.

In order to create a new trait I divided the length of the index finger by that of the ring finger to give a ratio for each individual. Sure enough, on average men had lower values of the finger ratio than women. The most important test, however, was whether finger ratio related to sperm measurements. I set my computer to look at sperm numbers, sperm speed, testosterone and oestrogen levels and the hormones that controlled the concentrations of sex hormones. Test after test showed that the finger ratio was a more powerful predictor than asymmetry of these important measures of male fertility. I was still unconvinced that the finger ratio was pointing to the foetal levels of hormones these patients had experienced in the womb. One test remained to be done – whether the ratio showed sex differences in very young children. My work on Liverpool and Jamaican children and adults clearly demonstrated that finger ratio shows similar sex differences across all age groups, from the very youngest boys and girls to the oldest men and women. I felt I had a good case for that rarest of things – an indicator of sex-hormone levels in the foetus, a kind of living fossil in the fingers which speaks to us of our exposure in the womb to these powerful and dangerous sex hormones.

That, then, was how I arrived at the link between sex and fingers, and an exciting journey it has been. My aim in this book is to explain what the finger ratio reveals to us about our fertility, behaviour and predisposition to disease.

I

A Tale of Two Fingers

The ways in which men and women differ reach into almost every aspect of our behaviour and health – verbal, mathematical, spatial, musical and other abilities; strength, running and swimming speed, jumping height and distance, handedness, throwing accuracy and distance; the prevalence of heart disease and the probability of heart attack, predisposition to most cancers, asthma, rheumatoid arthritis, autism, dyslexia, schizophrenia, neuroticism, psychoticism, hyper-activity, attention-deficit syndrome, stammering, migraine, depression and many tropical diseases. All these and more show different intensities of development, or are more common, manifest earlier or once manifest have different patterns or rates of progression in one sex compared to the other.[1,2] The differences stretch to the most apparently trivial things – for example, magnetic resonance imaging scans have shown that the rates of opening and closing the mouth in the watery world of the foetus are markedly different in boys and girls (girls 'mouth' more often than boys[3]). Sex differences span the whole range of traits, from those that are apparently trivial to those that may result in early death.

How do these differences arise, and why do some men show what we might call female-type traits, while some women seem rather masculine? It may sound bizarre, but our fingers, and more precisely our ring and index finger, can illuminate this debate. Our fingers provide us with evidence of how men and women differ, and how they are programmed before birth to show certain sex-related behaviour patterns and disease predispositions. In this book I aim to demonstrate that the early growth of our ring finger is sensitive to testosterone levels in the womb. Testosterone is the so-called 'male hormone', and the longer our ring finger the more 'masculine' we will turn out to be. Finger length is also dependent on body size, so we must compare the length of our ring finger with that of one of our other digits. The comparison is best done with our index finger, since I believe its early growth is dependent on the 'female hormone' oestrogen. The relative length of our ring and index fingers may speak to us of the balance of maleness and femaleness of our body and mind. This point can be illustrated by reference to the ring and index fingers of that most apparently masculine of men, Giacomo Casanova.

Casanova was an eighteenth-century musician, gambler, soldier, philosopher, writer, librarian, aspiring priest and, of course, womaniser. It is anecdotal but entirely unsurprising that we find from his diaries that he had a ring finger noticeably longer than his index finger.[4]

Casanova's adventures took him across Europe, and it is during his stay in Spain that we learn of the relative lengths of his fingers. At the time Casanova was enjoying the hospitality of the German neoclassical painter Anton Raphael Mengs. Today Mengs's work is often regarded as cold and contrived, but at the time he was seen by many as Europe's greatest living painter, and Goya was one of his many students.[5]

In the course of his stay Casanova's relations with his host were less than cordial. Mengs was openly scathing at what he saw as Casanova's neglect of his religious duties, and he even attempted to evict Casanova from his household after he failed to take the sacrament at Easter. Casanova meanwhile complained that Mengs was a lascivious, bad-tempered, jealous, avaricious drunkard who beat his children to excess. It is against this background of growing mutual dislike that Casanova recounts a dispute in which he drew Mengs's attention to the hand of the principal figure in one of his paintings, claiming that it was faulty because the ring finger was shorter than the index finger. Mengs assured him that this was the correct human condition, showing Casanova his own long index finger and short ring finger. Laughing, Casanova displayed his badge of prenatal masculinity, and asserted that he was sure that his ring finger, unlike Mengs's, was 'like that of all the children descended from Adam'. A wager of a hundred pistoles was made, and a hurried comparison with the ring fingers of the painter's servants showed the 'Casanova pattern' to be the more common. The incident ended with Mengs making light of his discomfiture with a jest concerning his own uniqueness.

Both were unaware that the 'Casanova pattern' was not universal but characteristic of men, marking out those who were masculinised before birth. No doubt this news would have caused Casanova much further glee, though this may have been tempered by the theory that a 'Casanova pattern' may recall a brutish ancestral stage through which humans have recently passed.

The 'Casanova pattern' and 'Mengs pattern' represent masculine and feminine prototypes which transcend sex in so far as they can be found among both males and females. However, clues to what the 'Casanova pattern' signifies can be gleaned from the observation that it is more common in men. Likewise the 'Mengs pattern' is found in many men, but overall it is characteristic of women as a group. These are generalisations relating to very large groups of people, but inevitably they invite curiosity regarding individual ratios, including our own. Does our ratio confirm or confound our understanding of our personality, abilities and drives?

Reports of associations between the finger ratio and sperm numbers, sex hormones, homosexuality, athletic ability and so on have appeared from time to time in the press. Before I discuss the way the trait may be measured I must make two cautionary points. The first is that the ratio is associated with tendencies such as poor verbal ability and high sperm count. The conclusions may be valid for large samples of people, but when we consider ourselves we go from the general to the particular. It must be borne in mind that at anecdotal level there are many exceptions to many rules. Do not regard your ratio as a definite indicator of a premature heart attack or infertility, or indeed of immunity to such things. Second, in order to know whether you have an unusual or unremarkable ratio it is necessary to know the ratio of others of the same sex, geographical group and race. Only then can you say for certain whether you are close to the average or close to the extremes.

Now for your ratio. Experts in the measurement of humans use steel vernier callipers with a digital readout which may be fed into a computer, and which measure to 0.01 mm for this task. For our purposes a ruler will suffice to get a rough measurement. Straighten your fingers and look at the palm of your hand. We are concerned with the second and fourth fingers (the first is the thumb, the second the index finger and the fourth the ring finger). At the base of your index and ring fingers there are creases. Your index finger is likely to have one such crease, the ring finger a band of creases. Select the crease closest to the palm, and choose a point on the crease that is midway across the base of the finger. Mark it with a pen if you wish. Measure from there to the tip of the finger and estimate to the nearest millimetre. Do this for both right and left hands. Now divide the length of the index finger by the length of the ring finger. Try this first for the right hand. You now have a ratio. If it is 1.00 the fingers are equal in length. If it is less than 1.00 the ring finger is longer than the index. A low value is more likely to be found towards the male end of the distribution. This is the type of ratio noted by Casanova in his hand. In the north-west of England the average ratio for white Caucasian males is 0.98; 0.92 would constitute an extreme ratio likely to be found in the hand of a male rather than a female. If your ratio is greater than 1.00 then your index finger is longer than your ring finger. In the north-west of England the average ratio for white Caucasian women is 1.00, and a ratio of 1.06 would constitute a high value likely to be found in the hand of a female rather than a male. Now repeat the process for the left hand. You now have your finger ratio for both hands. In order to check your measurement procedure try it on Figures 1.1 and 1.2 (pp. 6–7 and 8–9). These show a pair of male (1.1) and female (1.2) hands. The male is typical in that his ring finger is obviously longer than his index finger. In fact the finger ratio is of the 'Casanova-type' in that it is 0.92 for both hands. This means he has been exposed to high concentrations of prenatal testosterone compared to oestrogen. The female subject has a high, female-type ratio in the right hand (1.00) that speaks of high prenatal oestrogen and is very similar to that of the average white female ratio in the north-west of England. However, her left hand has a male-type ratio of 0.94. Such differences between right- and left-hand ratios are not unusual.

Although your right- and left-hand ratios are likely to be similar, men have a slight tendency for right-hand values to be lower than left, an example of a rather strange difference between the sexes. In men the male form of a sex-related trait is most obviously expressed on the right side of the body, and in women the female form is best expressed on the left. This extends even to the formation of the testes and the ovaries. In animals in general a very few individuals are born hermaphrodite, meaning that they have one testis and one ovary. Usually the testis is found on the 'male side', i.e. the right side, while the ovary is on the 'female side', i.e. the left side. Among individuals with normal sex organs men tend to have a larger right testis and women a larger left breast. Moreover, if we consider those tasks at which females excel, e.g. the use of language, then women with larger left breasts score more highly than those with larger right breasts. The reverse is found in men: those with larger right testes tend to score more highly in male-favouring tasks such as those involving spatial judgement.

The hands may be rich in such traits. Take fingerprints. Ridges of skin occur on the fingertips, palms, toes and soles of the feet. Counts of the ridges on the fingertips show that men tend to have higher numbers, and both sexes tend to have higher counts on the right compared to the left hand. However, in men this difference between right and left is more pronounced. When considering finger ratio the meaning of such differences is not yet entirely clear. There is some evidence to suggest that the right-hand ratio is more reliable as an indicator of such things as age at first heart attack and athletic ability in men, and that a lower ratio in the right hand may be an additional indicator of high testosterone. We need more evidence before we can be sure of these tendencies; for the present, it is best to use your right-hand measurement as the most

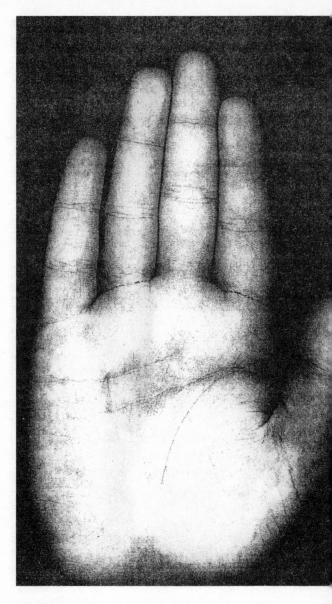

FIGURE 1.1
The hands of a white Caucasian male with creases marked. The 'life
finger length measurements are right hand (i) index finger 69.43mm
(ii) ring finger 75.86mm; right finger ratio **0.92**: left hand (i) index fi
68.30mm (ii) ring finger 74.46mm; left finger ratio **0.92**.

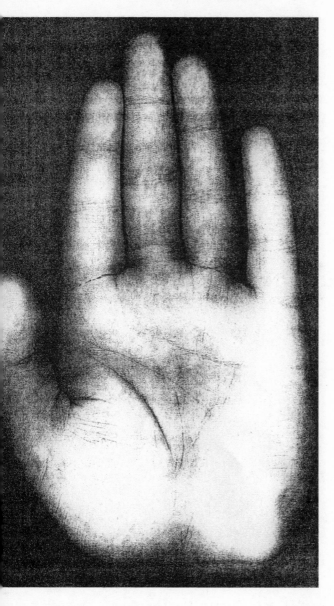

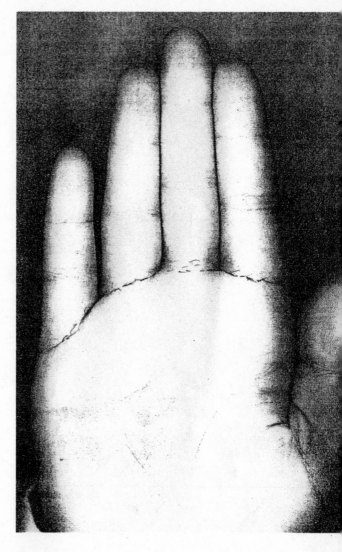

FIGURE 1.2
The hands of a white Caucasian female with creases marked. The 'life :
finger measurements are right hand (i) index finger 67.77mm (ii) ring
finger 67.67mm; right finger ratio **1.00**: left hand (i) index finger 65.79⊪
(ii) ring finger 70.25mm; left finger ratio **0.94**.

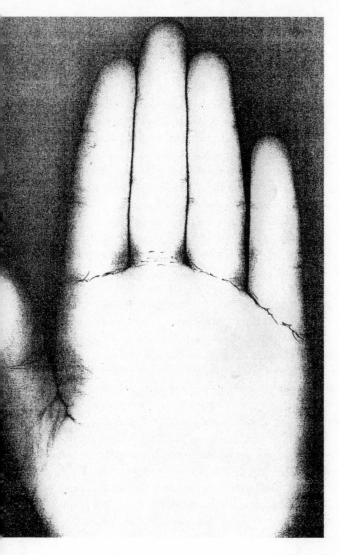

useful representation of your ratio. It is through this hand that a
whole suite of masculine or feminine traits may be revealed to be
dependent upon such early sex hormones as testosterone. In the
spirit of all that is related to Casanova, let us take as our first
example the formation of the penis.

What determines penile length? Are fingers related to the penis?
The simple answer to the first question is that testosterone produced by the male foetus causes the penis to differentiate into its
characteristic form of a root, a shaft and a complex structure at
its tip known as the glans penis.[6] Much of the penis is made up of
erectile tissue enclosed in three compartments, two placed side by
side along the upper part of the organ, the third found below.
The penis becomes erect when these compartments are gorged to
their maximum extent by blood. Studies have yielded average
flaccid lengths of between 9 and 11 cm, and average erect lengths of
between 13 and 15 cm.[7]

The penis forms under the influence of testosterone. Other
testosterone-related physical features such as height and foot size
are very weakly associated with penis length, so that very tall men
with large feet have a slight tendency to have larger-than-average
penises.[8] In addition, our sexual predilections may also relate to
testosterone and penis length, a number of studies concluding that
gay men appear to have longer penises than straight men.[9] However,
the most likely answer to the question of a penis predictor appears
to be that finger length is an indicator of penis length. The longer
one's fingers the longer one's penis. Evangelos Spyropoulos and his
colleagues in the Naval and Veterans' Hospital of Athens have considered the penis length of fifty-two young men in relation to their
height, weight, waist and hip circumference and the length of their
index finger,[10] recording the dimensions of the penis when it was
very gently stretched. These measurements were used to determine
the length of the shaft and the glans and the volume of the penis. All
these variables were related to the length of the index finger. Spyropoulos and his colleagues did not measure the remaining fingers, so
we cannot be sure of their relationship to penis length. My guess is
that they would have found the ring finger the strongest predictor,

and that long ring fingers in relation to index fingers would be associated with longer penises.[11]

I think the reason for the finger–penis link goes back to the colonisation of the land by vertebrates. Our amphibian ancestors in the coal swamps of the Carboniferous period were heavily armoured, slow-moving animals with four limbs, the number of toes varying from species to species. These primitive amphibians breathed air by means of lungs, but lived almost all their lives in water. Most importantly, in common with their modern relatives, the frogs, newts and salamanders, they probably reproduced in the water. Animals with this mode of life do not require a penis. Sperm is simply squirted directly onto eggs or produced in packets called spermatophores that are deposited in water and picked up by females. The penis is an evolutionary response to reproduction on the land because it is necessary for fertilisation within the female. In a similar way, the five-toed or pentadactyl limb is an evolutionary response to efficient locomotion on the land. Thus in an evolutionary and developmental sense toes and fingers have become intimately bound up with the reproductive system – not just the penis and its complex ducts in the male, but also parts of the female reproductive system, the vagina and uterus, which are necessary for internal fertilisation. It is not surprising, then, to learn that the genes which were acquired to control the formation of our fingers and toes are also the genes which are essential for the development of our complex reproductive system.[12] The penis and its ducts and the vagina and the uterus only make sense in a land animal that has efficient limbs, so fingers are intimately bound up with our sexuality, a connection which goes back many millions of years to our non-human ancestors. If this is so, is there any evidence in folklore that we have made the connection between sex and fingers?

The simple issues of whether the ring or the index finger is longer and the consequences of this for our anatomy and behaviour have been debated by scientists and non-scientists. The 'Casanova pattern' in the fingers is considered by some to be the mark of an ugly hand – an atavistic hand recalling brute instincts and behaviours, modelling the form of the fingers of our ape and monkey relatives.[13,14] Thus the 'beast' in us is represented by the ring finger while the 'beauty' resides in the index finger. This notion has led to suggestions that the feminised 'Mengs pattern' is of a purer type,[15] a hand which signifies emancipation from our primate ancestry.

In the arts and in palmistry vague connections with maleness, sexual attraction, creativity, musical ability and an atavistic or primitive nature have all been ascribed to a long ring finger. The most powerful example of this is the case of Franz Liszt. A cast of his right hand reveals a very long ring finger, with the middle section of the finger dominating its structure.[15]

Extreme variation in the ratio between the index and ring finger is readily noticeable. If long ring fingers are linked to musical ability through the influence of testosterone, it is not surprising that the association has been noted in the hands of very gifted musicians. It was Darwin[16] who first suggested that music is about sexual courtship. Good music may simply advertise testosterone-related traits such as high sperm counts and a healthy heart and vascular system. Not very romantic perhaps, but if a female wants such traits in her sons so they may make her many grandchildren it may pay her to listen intently.

The connection between ring fingers and reproduction is a theme which repeats itself across time and in many cultures. The fourth digit is known as the annulus or ring finger, because from Roman times it has been associated with the wearing of rings. A ring on this finger often indicated the marital or reproductive status of the wearer. In some cultures the ring finger is said to have a direct connection to the heart and to indicate the health of the heart by its length.[15] It seems that historically some connections between fingers, sex and disease have been made.

Science, however, has been slow to identify the importance of such connections. That there is a sex difference in the relative lengths of men's and women's ring and index fingers has been known for more than a century.[17,18] Compared to sex differences arising at puberty the finger ratio is modest in its size and visibility, and it has been neglected. However, it bears testimony to that most important of periods, our early development in the uterus. Let us begin to look at what science can tell us about finger ratios.

II
Fingers, Sex, Class and Ethnicity

Finger length has something important to say about how our brain, reproductive system and heart and blood vessels function. At first sight this notion might seem no more credible than palm reading and other diverting but probably dubious practices. I vividly recall describing the relationship between fingers and sexuality, fertility, athleticism, heart disease, breast cancer and schizophrenia to a Dutch science reporter. When I had finished there was a pause which was partly due to the language barrier but also to his incredulity. I could not see his face (it was a telephone conversation), but I could hear noises and movement as he marshalled his reactions to my claims and attempted to translate them into English. Finally he managed, 'It is too much!' I knew exactly what he meant. In a sense I also feel embarrassed by the number of important insights that I believe the finger ratio gives us.

As we have seen, men with particularly long ring fingers compared to their index fingers are highly masculinised: they have experienced high testosterone concentrations at the end of their first three months in the womb. Women with very long index fingers have experienced a feminising environment before their birth. Fingers also tell us that many men have experienced a feminising early environment and many women a masculinising environment before their birth.[1]

As we will see in the course of this book, the finger ratio says a lot more besides. As we might expect, our lifetime reproductive success may be related to the ratio. More surprisingly, as we move away from the traits that are explicitly related to sex, we find that our index and ring fingers give us clues to how our developing brain was moulded before birth. Indications that this is so come from relationships between the ratio and right- and left-handedness, verbal fluency, visuo-spatial abilities and the condition of autism.[2] Finger ratio also appears to be linked to ability at music[3] and sports such as football, athletics (especially running) and slalom skiing.[4] Related to questions of efficiency in athletic disciplines, which place demands on our cardiovascular system, the ratio may tell us something about the probability of heart attack and at what age this may occur.[2] Heart attacks are more common in men than in women. However, a major cause of mortality in women is breast cancer. There is evidence that the relative lengths of the index and ring finger tell us something about the likelihood of breast cancer and, if the tumour does appear, the age at which this likely.[5] Moving into the area of immune function, it is possible that the finger ratio alerts us to susceptibility to asthma, HIV and AIDS, and the probability of infection by important parasites such as the tropical roundworm that transmits river blindness.

This list of associations and insights may seem more than enough for two fingers. However, there are more. The relative lengths of the index and ring fingers may contain information related to our sexual orientation,[6] to class (or at least socio-economic level) and to our ethnic origin.[7] If this is true, finger ratio is a biological and medical dream – something which is easily and non-invasively measured and may be used as a screening tool for infertility, early heart attack, breast cancer, HIV/AIDS and parasitic infections, and which may provide insights into some of the processes involved in brain differentiation. We must exercise caution with sensitive topics such as sexual preference, class and race. However, liberal views of homosexual behaviour, class and race are not necessarily threatened by the realisation that prenatal conditions relate to such things, nor are extreme ideologies necessarily strengthened by such data. We must first consider the data and get the science right; if necessary the political polemics can then follow. First I will examine how we have recently come to the realisation that the hidden world of the foetus has a profound 'programming' effect on our biology.

Foetal physiology and biochemistry are very difficult to measure directly. For example, there are many ethical and methodological problems associated with determining hormone concentrations in human foetal blood. Difficulties in measuring the foetus often lead us to measure children and adults instead. Take, for example, the important problem of coronary heart disease (CHD), a major chronic disease affecting adults. It has been found that some human populations have a higher incidence of heart disease than others, and that these populations also have high incidences of smoking, obesity, high cholesterol and high blood pressure. To combat this, in the 1960s and 1970s many people took to dieting, jogging and aerobic exercise, with a consequent reduction in heart-attack rates. However, by the mid-1980s it was apparent that many lean and hungry non-smokers were still dying of premature heart attacks. In fact, lifestyle factors account for only about 25% of adult deaths from CHD. Evidence is mounting that biological 'programming' at the foetal stage is an important factor in the remainder.

In the late 1980s David Barker and his colleagues at Southampton University pointed out that such measures as birth weight, birth length and head circumference may be sensitive indicators of the foetal environment.[8] Indeed, reduced foetal growth rates are now known to be related to fatal CHD, particularly among men. Diabetes (a disease commonly associated with problems of the blood vessels) is also more common in adults who were underweight at birth, as is high blood pressure. Further work has shown that high cholesterol and elevated levels of clotting factors may also be associated with reduced foetal growth. This, then, is the biological programming model of heart disease. It suggests that many of the characteristics and function of our cardiovascular system are determined before birth. We can keep the machinery in good order by regular maintenance, eating sensibly and taking exercise, but there is a limit to how effective this may be.

What can affect the assembly of a good heart and vascular system? Poor nutrition, hormones, drugs (such as nicotine and alcohol) and even stress have been suggested as influencing the foetal heart. These represent a mixture of genetic and environ-

mental factors, so it is important to realise that biological programming does not necessarily involve genetic programming.
However, the essence of this model is that prenatal influences, often
during the early stages of foetal growth, may trigger major diseases
in adult life.

To an evolutionary biologist with an interest in sex differences
the most striking pattern associated with CHD is the excess of male
victims. Heart attacks are uncommon among pre-menopausal
women, and so mortality rates from CHD are two to three times
higher in men compared to women. This suggests that exposure to
sex hormones before birth may be a powerful influence on adult-
onset CHD. It is the central theme of this book that many diseases
show sex differences in their expression (e.g. in their age of first
occurrence) and progression. These sex differences are indicators of
prenatal effects of sex hormones such as testosterone and oestrogen.
The finger ratio also shows evidence of sex differences and is prenatal in origin. This may mean that it contains much important
information regarding conditions within the uterus. We need only
measure it to access that information.

Sex may be defined as the biological characteristics that dis-
tinguish between males and females.[9] These include differences in
groups of genes called chromosomes (human males have one X and
one Y chromosome and females have two Xs), the kind of sex cells
the individual produces (i.e. sperm or eggs), the types of duct which
lead from the testes or ovaries (in men the epididymis, vas deferens
and seminal vesicles, and in women the fallopian tubes and uterus),
and the external genitalia (the penis in the male and the clitoris,
labia majora and labia minora in females).[9]

During the first six weeks of development the foetus has the
potential to develop into a male or a female. Around six weeks the
beginnings of testicular or ovarian development appear. Within
the developing testis a substance is produced which inhibits the
growth of the female ducts, and then the testis begins to produce
testosterone. Testosterone is a fascinating hormone. It is made from
cholesterol, the substance that is often listed among risk factors for
heart attack. It is a relatively simple and small molecule, having

nineteen carbon atoms, and it may be readily converted to the eighteen-carbon-atom form of the 'female' hormone oestrogen. Synthesis of testosterone in the male foetus begins at eight weeks and peaks at thirteen weeks of pregnancy. Thereafter it slowly reduces in concentration, reaching low levels some months after birth. Testosterone is essential for the formation of the male ducts, the penis and the scrotum. In the absence of testosterone the foetus will develop female ducts and female external genitalia. The main source of foetal testosterone is the foetal testes, but some is produced by a pair of glands lying close to the kidney, the adrenal glands, and some by the ovaries. This means that both male and female foetuses are exposed to testosterone, but males generally have higher levels. There appears to be little or no movement of foetal testosterone across the placenta into the mother or from the bloodstream of the mother into the foetus. This is stopped by a powerful enzyme (aromatase) which converts testosterone to oestrogen. However, in rare cases women with defective aromatase and who are pregnant with a male child show the effects of the testosterone produced by their foetus. Their voice deepens and they grow facial hair. This effect disappears after the mother gives birth. In addition to being found in the placenta, aromatase is also present in foetal ovaries. The foetus can therefore make its own oestrogen from testosterone.

An examination of fifty-six human embryos and foetuses at various stages of development has shown that the relative lengths of the fingers are established remarkably early in development.[10] By the seventh week of life the foetus has established near-adult proportions of the fingers relative to each other and to the lengths of the bones making up the hand. There is then a relative elongation of the bones at the ends of the fingers followed by a return to adult proportions. The result is that by the thirteenth week of pregnancy adult proportions are reached in the hand. This is sufficiently early to allow for an effect of testosterone and oestrogen on finger growth. A study of the finger ratio in children and young adults has found that it differs between the sexes as early as two years and shows no noticeable change at puberty. Relative finger length appears to be established before birth, and the relative length of the index and ring finger is different between the sexes at an early age: the timing of this development suggests that its association with sperm production and testosterone levels is established before birth. Similarly, it was found that both women and men with female-type finger ratios had high levels of oestrogen as adults. Because the finger ratios are established early, this relationship may well have its origins in foetal development.

We may wonder whether there is any evidence that the ring and index fingers are especially sensitive or insensitive to testosterone and oestrogen. There is indirect evidence that the skin of the ring finger is sensitive to testosterone while the index finger may be unresponsive to this hormone. The growth of hair on the fingers of men and women has been the subject of study in such places as Europe, Nigeria, Namibia, South Africa, Tibet and the Solomon Islands.[11,12] Fingers have within them three bones,[13] the phalanges. Look at the backs of your fingers. The part of each finger closest to the hand proper corresponds to the phalanx that articulates with a bone within the hand itself. The back of this part of the finger is likely to be quite hairy. The middle part of the finger corresponds to the mid-phalanx, and this may have some hair on it. The end of the finger or terminal phalanx is unlikely to have hair. We are most interested in the mid-finger hair pattern. The existence and growth of mid-finger hair is dependent on testosterone, and in particular on its most biologically active form, dihydrotestosterone (DHT).[14] If all individuals and all fingers are equally responsive to DHT then we would expect to find approximately equal numbers of hairs on all fingers of all people. However, it is quite common to find individuals who do not have mid-finger hair. When mid-finger hair is present it is most luxuriant on the ring finger, sometimes found on the middle finger, rarely on the little finger and very rarely indeed on the index finger. I interpret this to mean that considering fingers two to five, the skin of the ring or fourth finger is most sensitive to DHT and that of the index or second finger least sensitive. If this sensitivity to testosterone and its derivates extends down to the tissue surrounding the soft cartilaginous phalanges of the foetal hand, it may translate into more rapid growth of cartilage, and therefore more rapid growth of the ring compared to the index finger.

Fingers may therefore provide a window into the important process of sexual differentiation in the womb. However, it is likely that the finger ratio and the distribution of mid-finger hair are not the only finger traits revealing prenatal environment. Ridges on the fingertips and on the palm of the hand are established before the nineteenth week of foetal development. Fingerprints with their arches, whorls and loops and their complex variants such as tented arches, radial and double loops were first described in the late nineteenth century.[15] The unique nature of each individual's fingerprint patterns may reflect levels of testosterone in the womb. Ridge counts show differences between right and left hands, often with higher numbers on the right hand. This difference is particularly strong for males. Ridge numbers are fixed early, so that sex differences in right and left hands may reflect prevailing concentrations of early testosterone. At present we have one study that shows a relationship between adult testosterone and patterns of ridge counts.[16] A reasonable interpretation of this result is that differences in testosterone between foetuses are preserved in differences in this hormone between adults. So, fingers are rich hunting grounds for sex differences which are fixed early and which may represent early concentrations of sex hormones. But why are fingers related to sex?

There may be a simple answer – the same genes control the development of the fingers and the sex organs. In 1997 Takashi Kondo and his colleagues published a paper entitled 'Of Fingers, Toes and Penises' in the journal *Nature*.[17] Although the paper contained little new data, its authors set out the evidence for a genetic link between fingers (and toes) and the reproductive system. The link lies in the large group of genes known as *Homeobox* or *Hox* genes. These are highly conserved and conservative genes – they have been around for a long time, are very important, difficult to change because of potential harmful effects, and similar *Hox* genes are found in very different organisms such as mice and humans. *Hox* genes control the formation of structures and organ systems including the head, chest, abdomen, limbs, and the urinary and reproductive systems. Some of these *Hox* genes control both the differentiation of the limbs and the urino-genital system. Why should a gene control the formation of such distinctly different things as a finger, a toe and a penis?

To address this question, let us go back to our distant ancestors, the lobe-finned fishes. Their present-day representatives, the lungfishes, live in water bodies that are prone to drying up. They have evolved lungs so that they are not dependent on gill-breathing in water. They have also acquired the ability to dig deep into the mud to avoid the consequences of temporary but long periods of little or no surface water. In such an environment lungs are of immediate use. However, they also constitute what is known as a pre-adaptation. They are potentially a platform for the development of an amphibious lifestyle.

Amphibians evolved to fill the freshwater niche. The early members of the group appear to have lived in rivers and lakes in a time of seasonal droughts. They were predators, feeding mainly on small fish. In the prevailing conditions the ability to walk from one stagnant and rapidly vanishing body of water to another, which may be deeper and therefore more likely to provide food, is an important trait. There was therefore a strong selective pressure to develop limbs. Given the invention of limbs the early tetrapods rapidly built up a diverse assemblage of land fauna. Food would be available on the banks of streams and pools in the form of stranded fish, and there would be an abundant supply of insects. But this mode of life means that another major evolutionary invention is favoured, an egg surrounded with an impermeable shell and which contains its own supply of water. Such an egg is independent of external water for its development. This is the amniote egg and it is characteristic of a new group, the reptiles. The stage is now set for the evolution of limbs that are truly adapted for terrestrial life. The early amphibians had heavy cumbersome limbs. After an initial experimental period they acquired the primitive five-toed (penta-dactyl) limb.[18] Why toes and why five of them is not known, but this basic plan has been retained ever since. The early amphibian limbs were not efficient load-bearing structures, and were not intended for a mainly terrestrial life. The reptiles, derived from the amphibians, had both efficient limbs and an amniote egg, which is impervious to water loss but also to the entry of sperm. Fertilisation has to occur internally, before the egg is surrounded by its shell. Internal fertilisation requires a complex series of ducts and usually a male organ which can inserted into the female. Hence new limbs, a penis (or at least some means of introducing sperm into females), complex reproductive ducts and an amniote egg are necessary to make a successful land vertebrate. *Hox* genes influence the development of fingers, toes, and penises, and so may owe their origin to the conquest of the land.

To the best of our knowledge, human toes do not show sex differences in the same way as the finger ratio. There are, however, interesting variations in relative toe length which may eventually be found to relate to sex. Toes are numbered in a similar way to fingers (big toe=1, next toe=2, and so on). About half of all Caucasians have toes 1 and 2 of about equal length and half have the second toe longer than the first. It would be of interest to know if having a big toe longer than the second toe is more common in one sex than the other. In the human foetus of eight to nine weeks the longest toe is the third, a pattern similar to relative finger length. In this the human foetus resembles adult apes. This may be explained by a tendency for patterns of individual development to mirror past evolutionary forms.

Finger ratio aids our understanding of sex differences, but does it shed any light on the factors influencing socio-economic status? Most of us have heard of Marx's belief in a conflict between a dominant and subordinate class.[19] This struggle, he argued, would be finally resolved when the former, the bourgeoisie, was displaced by the latter, the proletariat. He envisioned the revolutionary transfer of power to be more likely in industrial rather than agricultural societies, and claimed that socialist societies would be free of class conflict. This is a bold and inspiring theory that offers a clear prediction. The current opinion seems to be that he got it wrong. In most societies there are inequalities. In market-driven societies it is usual to find that the distribution of earnings exhibits a marked skew, a 'bell-shaped', so-called normal curve, with a long upper tail extended by the presence of a few very rich individuals. In such societies earnings are determined by a labour market, and an individual's position is dependent on ability to fill a particular occupation and, to an extent, on parental situation. In England and Wales about a third of all men born into a working-class family attain a middle-class occupation. Given stability in the number of middle-class roles and roughly equal reproductive rates between the classes, this suggests that a similar proportion of middle-class children end up filling working-class occupations.[19] What factors other than parental rank determine an individual's occupational role? Maybe we are all born with the abilities and enthusiasms that would enable us to fill any role – in effect a blank slate that may be filled with the detailed writings of people who are immediately around us and with the broad brush strokes of society. In the 1920s and 30s the Behaviourists believed that individuals could be conditioned by reward and punishment into any role, that any healthy child could be trained to become a doctor, lawyer, scientist, tradesman or criminal. In a similar vein, sociologists such as Emile Durkheim believed that people were moulded into roles by a society's 'collective consciousness'.

Others have argued that occupational roles are often filled by those whose innate abilities fit them to such roles. Caution must be exercised here. This is not to suggest that we are hard-wired to choose certain jobs, rather that tendencies springing from our biology make it likely that we find certain jobs easy or pleasurable to perform. This stance is seen as controversial, particularly when applied to the issue of why the proportion of men and women varies across occupations. However, consider the results of a recent online survey conducted through the BBC's science and nature websites. There were 255,000 respondents from more than 150 countries. All were asked to give their occupation (from a choice of twenty-five jobs) and to measure their second and fourth fingers. The proportions of men and women varied across occupations. For example, the workforce in engineering, skilled labour and IT was predominantly male (greater than 80%), while homemakers, administrators and those in healthcare jobs such as nursing and occupational therapy were predominantly female (greater than 70%). An inspection of the finger ratios of the women in the survey showed more masculine ratios among female engineers and those in skilled labour and IT, and the most feminine ratios among homemakers, administrators and healthcare professionals. I interpret this to mean that prenatal influence is subtle, but it is nevertheless one factor involved in occupational roles.

One possible influence on occupational choice which is hotly debated is 'intelligence'. In their controversial book *The Bell Curve* Richard Herrnstein and Charles Murray argue that twentieth-century American society has seen an increase in stratification according to 'cognitive ability'.[20] Their measure of such ability is the Intelligence Quotient (IQ), a concept developed in the early twentieth century to express an individual's intellectual performance relative to a given sample or population. Its significance is disputed, but it has been demonstrated to be a predictor of occupation. Lawyers, college teachers, accountants, engineers, architects and scientists have an average IQ of about 120, a substantially higher score than the population average of 100. This is not to say that these professionals necessarily need a high IQ to do their job, rather that intense selection for high IQ has concentrated an increasingly high proportion of cognitively gifted individuals within these jobs. What has this to do with genetics? Herrnstein and Murray point out that there is evidence that IQ is at least in part inherited, and that if success requires high IQ, and earnings depend on success, then position in society is associated in some way with inherited differences.

IQ is not strongly related to gender, as those abilities that are markedly different between men and women are excluded from IQ tests. It is therefore unlikely that the finger ratio is correlated in a simple way with IQ. However, foetal concentrations of testosterone may have an effect on adult occupational roles, independently of IQ. Lower socio-economic occupational roles are associated with hard physical labour. There is accumulating evidence that testosterone enhances the efficiency of the cardiovascular system in men.[21–23] This may come as a surprise to many, given that all things male, such as higher testosterone and higher job-related stress, are seen as being associated with early and often fatal onset of heart disease. Oestrogen, on the other hand, is seen as a protective against heart disease because pre-menopausal women rarely have heart attacks. The reality is that both testosterone and oestrogen are important for healthy vascular systems. The former is protective against vascular deterioration in men, while the latter has a similar protective function in women. Men who have had heart attacks tend to have lower levels of testosterone than healthy men who are of similar age and body weight. Testosterone may enlarge the blood vessels, resulting in less resistance to blood flow, and may also protect against thickening of the arteries. In contrast, high oestrogen levels may be related to increased rates of heart attacks in young men. Given testosterone's association with an efficient heart and vascular system in males, it is unsurprising that high testosterone has been identified with male socio-economic roles that place stress on repetitive physical labour.[24]

Associations are notoriously difficult to interpret, but the relationship between adult testosterone and low socio-economic level has been tested, in a study of 272 men living in the north-west of England who had experienced a heart attack and/or angina pectoris.[25] The average age of the study group was sixty-three years, with a range of twenty-nine to ninety years. The socio-economic level of the participants was calculated in the form of Townsend Deprivation Scores, which are derived from four estimates of the prevalence of deprivation within postal districts – the frequency of unemployment, non-ownership of a car, occupation of rented accommodation and living in an overcrowded household. The scores in this sample ranged from –5 (the most affluent) to 10 (the least affluent). High deprivation was associated with two variables in these men, a low finger ratio and youth. It is possible to examine the relationship between finger ratio and deprivation with the effect of age removed and the relationship of age with deprivation with the effect of finger ratio removed. Both finger ratio and age independently predicted socio-economic status; indeed, the relationship between finger ratio and deprivation is stronger when the influence of age is removed. Why is a 'Casanova-type' finger ratio associated with high deprivation? Suppose this trait is related to high testosterone levels before birth and, as a consequence, an efficient cardiovascular system. Such men might be able to produce high testosterone concentrations as adults, may therefore find that demanding physical work is easy and even enjoyable, and will gravitate towards manual jobs. We must be cautious with this notion, since the relationship between finger ratio and deprivation is weak, and there has been only one study to date. Even if the association is found to be general, it is obvious that prenatal testosterone is not the only factor which may be predictive of socio-economic level.

Let us speculate further about early testosterone and class. There is now evidence to suggest that the finger ratio is inherited.[7] This may mean that prenatal testosterone levels and, by extension, an ability for hard and sustained physical labour may also be inherited. Note that abilities associated with high testosterone and with high IQ are likely to have been valued differently in different societies at different times. Herrnstein and Murray stress that the strong association between IQ and the 'top of the American labor market' has arisen in the latter half of the twentieth century. In societies less strongly stratified for cognitive abilities, men with an ability to work hard and for long hours at physically exhausting tasks may enjoy high status and earnings. This is another way of saying that selection forces acting on genes for testosterone production and genes for IQ may vary in strength with time and across societies. This prompts us to look at differences in finger ratio across geographically and ethnically distinct populations.

There are differences in adult testosterone concentrations between human populations. For example, a comparison of hormone levels of men of the !Kung San of the Kalahari Desert and Kavango men living along the boundary between Namibia and Angola has shown large differences in testosterone.[14] The San are small in stature, have weak facial hair growth beyond the upper lip and the sides of the chin and have nearly hairless bodies. Kavango men are taller and more robust, have strong, uniform beard growth and well-developed hair growth on the chest, abdomen, pubic area, arms and legs. Not surprisingly Kavango men have higher testosterone, higher concentrations of DHT, and higher levels of unbound and therefore active testosterone than the San. Differences in total testosterone have also been noted between black and Caucasian populations. A Californian sample of African-American and Caucasian students showed 15% higher testosterone concentrations in the former than the latter.[26] This was after socio-economic status and lifestyle factors were taken into account. However, we are more concerned with the idea that testosterone levels before birth are much more important than concentrations in adults. Indeed, there may be little correlation between the two. It is very difficult to measure testosterone levels in foetuses. Therefore little is known about prenatal differences in sex hormones between ethnic groups.

Estimates of population values for finger ratios are now available for samples from England, Spain, Germany, Poland, Finland, Hungary (both ethnic Hungarians and Hungarian Gypsies), southern India, South Africa (Zulus) and rural Jamaica.[7] It is clear from these samples that the differences between the sexes are more or less constant across populations, with males tending to have longer fourth fingers relative to their second than females. The average of this difference is about 0.02. For example, among white Caucasians in north-west England the mean male ratio is 0.98 and that of females 1.00. In the black Jamaican sample the mean male ratio is 0.93 and the female ratio 0.95. This illustrates real differences between the sexes and between the ethnic groups, although it is in the ethnic comparisons that we often see the most marked differences. Figure 2.1 shows this variation. Evidence of high ratios was found in the samples from central Poland in and around the city of Poznan, the north-west of England and the Granada area of Spain. Intermediate values of finger ratio were found in individuals from the area around Pecs in southern Hungary (both ethnic Hungarians and Gypsies), the cities of Bielefeld, Hamburg, Hannover, Kassel and Kiel in Germany, and tribal Sugali and Yanadi groups from

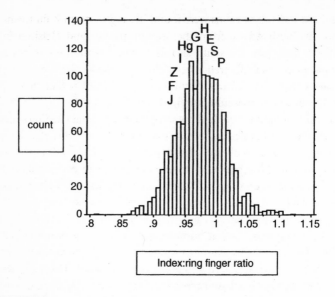

FIGURE 2.1
In this sample of 1,516 men and women from ten countries, the average index:ring finger ratio is 0.97 and ranges from about 0.87 to 1.12. The distribution of the finger ratio is in the form of a 'bell-shaped' or 'normal' curve, which is typical of biological variables. The average ratio per country varies across this distribution and is indicated by the position of the country codes. The highest average ratios are found in the samples from Poland (P), Spain (S) and England (E), in a band of ratios from 1.00 to 0.98. These populations (both men and women) tend to have quite long index fingers in relation to their ring fingers. They may well have been exposed to lower testosterone concentrations before birth than the subjects from other countries in the overall sample. An intermediate group (average ratios from 0.97 to 0.96) comprises subjects from Hungary (H=ethnic Hungarians and Hg=Hungarian Gypsies), Germany (G) and India (I). These subjects may have had higher prenatal testosterone than the Polish, Spanish and English participants, but lower than those of the next group. The group with the longest ring fingers (average ratio 0.95 to 0.93) comprise Zulu, Finnish and Jamaican participants. They are likely to have experienced high testosterone concentrations before birth.

southern India. Low ratios were found in samples from Zulu townships in South Africa, from the area in and around Helsinki in
Finland, and from the south side of Jamaica. Note there is not a
simple white/black difference here. However, the available evidence
suggests that black groups may be more uniform in that they all
tend to show low, masculinised finger ratios.

Recently unpublished data for samples from China (Beijing) and
Japan (Tokyo) have become available. Perhaps surprisingly, these
samples show masculinised finger ratios. Chinese and Japanese
men do not show markedly testosteronised features in pubertal
characteristics such as height, body hair and jaw size. Nevertheless,
the available evidence suggests high prenatal testosterone is common among east-Asian groups.

These data indicate that human populations show very marked
differences in their exposure to testosterone and oestrogen before
birth or in their sensitivity to early sex hormones. However, we
must be cautious here. We do not have direct data on ethnic differences in sex hormones before birth. Is it really the case, as suggested
by the finger-ratio data, that English men are exposed to lower
testosterone concentrations before birth than Jamaican women?
Before we can fully understand the variation in the finger ratio
across ethnic groups we need more data on foetal concentrations of
sex hormones.

Strong ethnic differences in finger ratio are also apparent in
children. Figure 2.2 shows the finger ratios of 1,010 children aged
from five to eleven years from four ethnic backgrounds – English,
Han and Uygur Chinese and Jamaican. The English sample is of
white Caucasian children from the north-west of England. The
Jamaican sample is of black rural Jamaican children. These two
samples represent the extremes of the distribution: the average
finger ratio for the English boys and girls was 0.99 and for the
Jamaican children 0.93. Between these two extremes of low and
high masculinisation are the Chinese participants, residents of the
North-West Province of China. The Han are an oriental people,
the most numerous race in China and the country's ruling group.
In the North-West Province they form an influential but minority

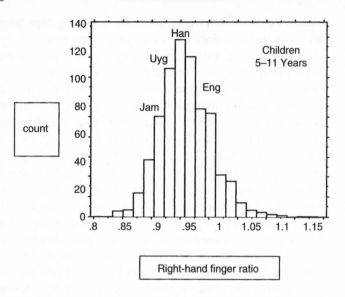

FIGURE 2.2
The finger ratios in four samples of male and female children aged five to eleven years. The subjects were white Caucasians from the north-west of England (Eng), Han Chinese from the North-West Province of China (Han), Uygur Chinese also from the North-West Province of China (Uyg) and black Jamaican children from a rural population on the south side of the island (Jam). The average finger ratio per population is indicated on the figure: 0.99 for the English sample, 0.95 for the Han, 0.94 for the Uygur and 0.93 for the Jamaican sample.

group. The Uygur, the most numerous of the ethnic groups of the North-West Province, have Caucasian features, dark skin, hair and eyes, a language that is very similar to Turkish and a script that is Arabic. Relations between the two groups have not always been amicable. Average finger ratios show higher, more oestrogenised ratios in the Han (0.95) and lower, more testosteronised ratios in the Uygur (0.94). Within the large range of population averages (from 0.93 to 0.99) there were also small but real differences between the sexes. These data indicate that sex and ethnic differences in finger ratio are found in children and therefore arise before puberty. But what happens to finger ratios when groups migrate and two or more ethnicities are found in the same area?

Where ethnically distinct groups live together but maintain a degree of reproductive isolation they often show differences in reproductive rates and socio-economic level. For example, in a Hungarian sample consisting of ethnic Hungarians and Hungarian Gypsies the latter showed higher deprivation rates than the former.[7] Some Gypsies lived in the poorer areas of the towns, while others were found in Gypsy villages and were often desperately poor. In this sample the average number of children per Gypsy adult was 3.48 and per ethnic Hungarian 2.41. The Gypsies also had lower finger ratios than the ethnic Hungarians, 0.96 versus 0.97. Larger families require a division of parental investment into smaller amounts per child, magnifying the difference in deprivation already existing between Gypsy and non-Gypsy children. I think it is unlikely to be a coincidence that Hungarian Gypsies and Uygur Chinese have more testosteronised finger ratios than the economically dominant groups in their area, and would suggest that in general economically advantaged groups will show more feminised ratios than disadvantaged groups living in the same area. For example, in the UK white Caucasians are likely to have higher finger ratios than people of Asian origin who in turn are likely to have higher finger ratios than black people. In the US my suspicion is that whites will have higher ratios than Hispanics who will have higher ratios than African-Americans.

The finger ratio, then, shows sex, socio-economic and ethnic differences. It is now time to examine these issues of sex, class and ethnicity in detail.

III

Fingers and Sex Hormones

The conclusions I draw throughout this book rest on the assertion that the finger ratio reflects the concentrations of testosterone and oestrogen we have been exposed to while in the womb. Such a statement needs to be supported, and in this chapter I wish to discuss the available evidence for the link between early sex hormones and growth of the 'Casanova-type' and 'feminine-type' of finger ratio.

It is my belief that the differences in finger ratio between males and females result from the production of testosterone by the testes. The male foetus starts to produce testosterone from his own testes by about week eight. This production peaks at about week thirteen.[1] Small amounts are also made by the adrenal glands, and in females by the ovaries. However, the possession of testes means that males have the edge on testosterone production, and therefore on the whole have lower finger ratios than females.

Sex and ethnic differences in finger ratio are important, but of equal interest are the differences in finger ratio between individuals of the same sex and ethnicity. Finger ratios run in families, and there is a weak tendency for sexual partners to have similar finger ratios. Therefore, the children of masculinised-ratio parents tend also to have low ratios, as the children of parents with feminised ratios tend to have high values of finger ratios. This applies for boys and girls, and hence there is much overlap between the sexes.

Low ratios may be explained not only by exposure to high testosterone in the womb, but also by sensitivity to testosterone through the structure of the androgen receptor gene. Similarly, feminine ratios may be related to high prenatal oestrogen and/or high sensitivity to oestrogen through the structure of the oestrogen receptor gene.

Since I suggested in 1998 that finger ratio was dependent on the balance of testosterone and oestrogen experienced in the womb at about the end of the first three months of pregnancy, I have seen accumulating evidence that these suggestions were correct.[2] However, much uncertainty remains. For example, factors other than sex hormones may be important in the formation of the finger ratio. One such is the hormone cortisol, which is secreted as a response to stress and is associated with feelings of fearfulness and uncertainty. Cortisol levels tend to be high when testosterone is low.[3] Hence high levels of testosterone, with its traditional association with assertiveness and competitiveness, are incompatible with the feelings of fearfulness correlated with cortisol.

Whether finger ratio is associated with maternal stress and cortisol production during pregnancy is not known, but at present testosterone and oestrogen appear to be the best candidates for the formative role in finger-ratio development. I say this because of the patterns of finger ratios with regard to women's body shape, in children who produce very large amounts of testosterone while in the womb, and in relation to genetic variation in sensitivity to testosterone.

Women's body shape is influenced by levels of oestrogen and testosterone. Compared to men, women have an 'hourglass' shape, with a small waist and large hips. This body shape is measured as the ratio between the circumference of the waist and hips, or waist-to-hip ratio (WHR).[4] The WHR of human females changes with their reproductive status. In children of both sexes an average WHR of between 0.8 and 0.9 is common, i.e. the circumference of the waist is about 80% to 90% that of the hip circumference. However, at puberty females produce large amounts of oestrogen, which encourage fat deposition on the hips and thighs but not on the stomach. This, then, is the 'hourglass' or gynoid figure, which in young fertile females is associated with a WHR of about 0.7. Testosterone also influences fat deposition, but around the waist.

In young men this results in a 'tubular' figure with an average WHR of about 0.9 to 1.0. These average WHRs may change with age and even with short-term changes in hormone levels. In women WHR increases after the menopause, when oestrogen levels reduce, but this may be delayed by the use of HRT. Women's WHR may even change across their cycle, with low WHR being found at the mid-point or fertile point of the cycle.[5]

To say that the average WHR of young and fertile women is about 0.7 does not give the correct 'feel' for the variation around this figure. Many women have a higher, more male-like WHR, giving them a tubular type of body. As one might expect, this form of WHR is associated with quite low oestrogen and high testosterone; all of this variation in adult levels of sex hormones and body form has implications for fertility, health and attractiveness. We owe much of our knowledge of WHR and partner preferences to the work of Devendra Singh of the University of Texas, whose work has shown that men tend to prefer women of a low WHR (about 0.7), as such a partner tends to be young, fertile and healthy.[6,7]

WHR and finger ratio both indicate sex differences that are dependent on the relative amounts of testosterone and oestrogen, and both show evidence of genetic influence. However, the fundamental difference between the traits is that finger ratio is determined by prenatal sex hormones and WHR by adult sex hormones. Hence the direct relationships between finger ratio and WHR are probably real but weak – women with high WHR tend to have low finger ratios, but the association is not very strong. Nevertheless, a recent study of ninety-five Jamaican mothers and their children has shown an association between the WHR of the mothers and the finger ratios of their children. Curvaceous women pass on genes for high oestrogen and low testosterone to their daughters and sons who therefore have high finger ratios. Tubular women pass on genes for low oestrogen and high testosterone and have daughters and sons with low finger ratios.

Although variation in WHR falls within the normal span of human differences, excess testosterone may be the result of disease.

Children with congenital adrenal hyperplasia (CAH) have enlarged adrenal glands at birth. The adrenal glands are small, walnut-sized organs closely associated with the kidneys. One important function of these glands is the production of cortisol; CAH children are unable to make enough cortisol, and their adrenal glands enlarge in an effort to compensate for the shortfall. This enlargement leads to the overproduction of other substances synthesised by the adrenals, including male hormones similar in structure and effect to testosterone. CAH is easily treatable with cortisol-like drugs, but some of the effects of early excess male hormones remain with the children. Thus CAH girls tend to have masculinised genitalia, with an enlarged, penis-like clitoris, often show masculinised, rough-and-tumble play behaviour, and after puberty may display increased bisexual and homosexual fantasies and behaviour. These male-type traits are an indication that early exposure to testosterone-type hormones may have an effect on the developing brain.[1]

Do CAH children tend to have masculinised finger ratios in addition to masculinised genitalia? Aysenur Okten and his colleagues at the Karadeniz Technical University, Turkey, have compared the finger ratios of seventeen female and nine male CAH children with finger ratios from a control group of 104 healthy children, half of whom were boys and half girls. The CAH and healthy children were matched for age. For both boys and girls, they found that CAH children had lower finger ratios than healthy children. Moreover, the finger ratios of CAH girls were indistinguishable from those of the healthy boys. Okten and his colleagues also measured finger-bone length from X-rays, but found no differences between the sexes or between CAH and healthy individuals. It may be that sex differences in finger-bone ratios are weak.[8]

A similarly low finger ratio in CAH children has been observed by Marc Breedlove of Michigan State University and his colleagues at the University of California and City University of London. This study reported finger ratios from thirteen females with CAH matched with forty-four healthy females, and sixteen males with CAH matched with twenty-eight healthy males. In both males and females the finger ratios of CAH individuals were lower than those found in their healthy controls. One useful feature of this report was that a small subset of CAH boys was matched against their healthy brothers. This careful matching showed lower finger ratios in the right and left hands of the former compared to the latter.[9] I conclude from these studies that CAH children tend to have masculinised finger ratios as a result of exposure to high prenatal levels of testosterone-like hormones. There are other possibilities, however: for example, masculine digit ratios may result from low cortisol in the foetus rather than high levels of masculinising hormones.

These findings do not demonstrate a direct link between finger ratio and sex hormones, but they constitute persuasive circumstantial evidence. At present there has been only one study that provides evidence for a direct relationship between finger ratio and testosterone, and this is concerned with sensitivity to testosterone rather than amount.

Testosterone influences the expression of many genes which in turn influence other genes, and so it has an immensely influential 'cascade' effect on the body. However, before this cascade can be activated testosterone must first bind to a receptor molecule, the androgen receptor; the hormone-receptor complex then exerts its effect on the genes. The androgen receptor is therefore very important, as its structure determines our sensitivity to testosterone. The androgen receptor is a protein which, in common with all other proteins, is made up of building blocks known as amino acids. One such amino acid, glutamine, is found in large numbers in a particular part of the receptor molecule. It is the length of this glutamine chain that is important in determining our sensitivity to testosterone.[10] In general the average number of glutamines per receptor molecule is around twenty to twenty-two, although the normal range is from about eleven to thirty.[11] If you have eleven glutamines your testosterone-receptor complex is very effective in switching on the testosterone cascade, but as the number of glutamines increases the sensitivity to testosterone reduces, so that an individual with thirty glutamines will have mild testosterone insensitivity. Very long glutamine chains can lead to profound problems, so that forty or more is associated with complete testosterone insensitivity, leading to infertility, muscle weakness and neuro-degeneration (Kennedy's disease). It is the normal variation, however, that concerns us here.

The various forms of the androgen receptor have important consequences for our health and behaviour. For example, African-American men have shorter glutamine chains (high sensitivity to testosterone) than white American men.[12] Short glutamine chains are associated with an increased susceptibility to prostate cancer,[13] and this may in part explain why the incidence of prostate cancer is higher in African-Americans than in white Americans. Other health-related associations with short glutamine-chain length include protection against breast cancer in women, but susceptibility in men to some forms of liver cancer, kidney stones, rheumatoid arthritis and ankylosing spondylitis (a progressive 'fusion' condition of the spine).[14] Glutamine-chain length also has effects on fertility. Men with high numbers of glutamine units have lowered sperm counts,[15] while women with short glutamine chains have elevated frequencies of fertility-lowering conditions such as polycystic ovarian syndrome and spontaneous abortion of female foetuses.[16] Short-chain glutamine receptors have also been associated with certain types of behaviour more common in boys than in girls (the so-called naughty boy syndrome), including attention deficit hyperactivity disorder, related conduct disorders and oppositional defiant disorder.[17]

The associations between androgen receptor structure and health and fertility have some overlap with similar associations between finger ratio and health-related traits. Low numbers of glutamines and a 'Casanova-type' finger ratio are both related to high sperm counts, and two studies have thus far found that men with a 'Casanova-type' finger ratio tend to have more children than men with high finger ratio.[18,19] This is a surprising finding given that fertility is only one factor in determining family size. For example, the fertility of your partner, desired family size and the availability and use of contraception will tend to obscure relationships between your own fertility and family size. Nevertheless, finger ratio turns out to be related to family size in such disparate societies as the north-west of England, with its small family units and wide use of contraceptives, and Zulu townships near Durban with large family size and widespread lack of effective contraception. A relationship between fertility and finger ratio has also been suggested in men who have few or no sperm in their ejaculate (azoospermic males). One common cause of azoospermy is the unsuccessful reversal of vasectomy, which may render a previously fertile man azoospermic. Other causes are the absence of the tubes that conduct the sperm from the testes to the penis, chemotherapy, and then a range of undiagnosed reasons. Sperm may be retrieved from azoospermic men by testicular biopsy, and removal of a single developing sperm (a spermatozoon) from a cell by use of a fine glass needle. The spermatozoon may then be injected directly into the egg by a technique known as intracytoplasmic sperm injection or ICSI. The technique of ICSI was first described in 1992 by Palermo and his colleagues,[20] and it proved so effective that by 2000 there were 78 licensed centres for ICSI in the UK alone.[21] However, successful sperm retrieval is not always possible, and it would be valuable to be able to predict which patients would best benefit from this technique. The finger ratios may be helpful in this context, and Simon Wood and his colleagues at Liverpool University have recently shown that men with a low, masculinised finger ratio are more likely to have successful sperm retrieval than those azoospermic men with high, feminised finger ratios.[22]

The sex ratio of families may also be associated with both testosterone sensitivity and the finger ratio. In women with short glutamine-chain length there is a tendency for spontaneous abortion of female foetuses which may result in an excess of male births.[23] In a similar way women with low finger ratios tend to give birth to more boys than girls.[24]

Risk factors in breast cancer are also related to both glutamine-chain length in the androgen receptor and variation in finger ratio. As we will see in Chapter V (pp. 56–70),women who are sensitive to testosterone are less prone to breast cancer.[25] This may be because the action of testosterone somehow opposes that of oestrogen, which is a known risk factor for breast cancer. Interestingly, another gene that confers protection against breast cancer, BRCA2, apparently exerts some of its effect through increasing the expression of the androgen receptor gene. Following this logic, it would appear that a low finger ratio may exert a protective effect against breast cancer. This has been found in a study of 118 Liverpool women with breast cancer. Although the cancer patients did not have significantly different finger ratios than healthy women, finger ratio was predictive of age at breast cancer, with high-ratio women developing the disease earlier than women with low ratios.[26]

If low numbers of glutamine units and low finger ratio are both associated with male fertility and protection against breast cancer it may be that finger ratio is correlated with the structure of the androgen receptor. This has been found in one sample of fifty men recruited from Liverpool University students, academics and Liverpool track athletes. In this sample the average number of glutamine units in the androgen receptor was twenty-one and the range was eighteen to twenty-eight. An examination of the association between finger ratio and glutamine number revealed that right-hand (but not left-hand) finger ratio was associated with glutamine-chain length. Those men with a 'Casanova-type' ratio in their right hand were more sensitive to testosterone than men with a high finger ratio. This is the first report of a direct link between finger ratio and testosterone. In this case it was sensitivity to testosterone rather than amount of hormone that was related to finger ratio.[27]

Much research remains to be done on how the finger ratio forms. We need to define the precise point at which the finger ratio becomes different between the sexes, and identify the factors, be they hormones, environment or genes, which determine finger ratios. This will enable us to understand what the finger ratio is telling us when we use it to study the origin of complex behavioural characteristics, and may even give us some clues about the formation of personality, a subject to which I will now turn.

IV

Fingers and Personality

We are all unique. This may be seen by a glance at the variation in our physical characteristics, but it is most intensely expressed in our behaviour. Consider for a moment your own personality. Are you conscientious? Do you show tendencies towards excessive emotional sensitivity or neuroticism? How do you score on such things as extroversion, openness and agreeableness? A surprising amount of this and other behavioural variation shows sex differences.[1]

Compared to females, males tend to be less neurotic and less intuitive when guessing the emotional state of others.[2] Males tend to display large muscle movements and physical aggression, but are also prone to hyperactivity, short attention span and language problems such as dyslexia, poor verbal fluency and stammering. However, high ability in spatial judgement, problem-solving, competitive sport, some aspects of mathematics, and writing and performing music are more likely to be found in males than in females.[3]

The source of these sex differences may lie in genes, early sex-hormone exposure and socialisation. All three are likely to be important, but I am primarily interested here in variation within the sexes. Females are less likely to be physically aggressive than males, but some females are more aggressive than others. Males are less likely to show excessive emotional sensitivity than females, but some males are more neurotic than others. Some of the variation in our behavioural tendencies and our personalities can be explained by the pattern of early exposure to sex hormones.

A 'Casanova-type' finger ratio is not shared by all men, nor is a feminised finger ratio found in all women. In order to see the influence of early sex hormones on our personality we must look at our finger ratios.

First I will consider the factors which influence behaviour in children under five years. My own recollection of pre-school years was of an intense physical and emotional period, a feeling confirmed when I observed this formative period as a parent. It was a time of fragile friendships, alliances against common enemies and great rivalries, of 'rough-and-tumble play' and frequent verbal and physical aggression. We played until we were exhausted, rested and then played again, and all this occurred against a backdrop of few self-imposed norms. Controls on behaviour during this time came from parents and other adults.

There has to date been one study of finger ratio and behaviour in children aged from two to five years. Justin Williams and his colleagues in the Department of Child Health, University of Dundee, measured the finger ratios of 108 pre-school boys and eighty-eight pre-school girls from ten local nurseries.[4] As expected, the average finger ratio of the boys was lower than that of the girls (at 0.95 and 0.96 respectively), but there was much overlap. Parents and teachers completed questionnaires concerning 'social difficulties' and 'social cognition' behaviours shown by the children. The 'social difficulties' questionnaire contained scales for measuring five important behavioural items:

(i) hyperactivity – on this scale children scored highly if they tended to be easily distracted, restless, with constant fidgeting or squirming

(ii) emotional items – high scores on this scale indicated children who were often unhappy, worried, tearful and nervous

(iii) peer problems – here a high score indicated a lonely, solitary child who was often picked on by bullies

(iv) conduct problems – high-scoring children often had temper tantrums, fights with other children and were argumentative with adults

(v) prosocial items – this scale was marked positively, high scores
     indicating children who were reported to be kind, helpful and
     considerate with others of their age group

The 'social cognition' questionnaire provided further information
on the children's awareness of the feelings of others, their ability to
judge body language, their social skills and responsiveness to com-
mands. The questions were marked negatively, high scores indicating a child with poor social cognition. There were sex differences
in some of the items, with boys scoring higher for hyperactivity,
conduct and peer problems and girls showing more prosocial
behaviour. This finding was consistent with previous work.[5]
   We might expect that a low finger ratio, with its association
with prenatal testosterone, would signal a tendency towards male-
like disordered behaviour such as hyperactivity, while high finger
ratios would be associated with such traits as emotional
behaviour. In part these expectations were fulfilled, but the
strongest associations were found in girls rather than boys. The
most masculine right-hand ratios in girls were associated with
hyperactivity, poor social cognition, low scores for prosocial items
and peer problems. Of these four associations two, hyperactivity
and poor social cognition, were particularly strong and therefore
likely to be genuine. Pre-school girls with low ratios have a tendency to be easily distracted and restless; their social skills may be
minimal, they may be very demanding of people's time and they
may lack awareness of people's feelings and fail to pick up on
body language. These associations are only tendencies, but they do
suggest that such behavioural traits in girls are at least in part the
result of high prenatal testosterone. In contrast, the male sample
produced only one strong association. Boys with high ratios
associated with elevated prenatal oestrogen were reported to score
highly on emotion, with a tendency to complain of headaches, to
be nervous or clingy in new situations and to have many fears and
worries.
   This sample appears to be telling us that variation in prenatal
sex hormones has a greater effect on female behaviour among

pre-school children. The explanation may be that while increasing concentrations of testosterone result in an increasing tendency to such behaviour as hyperactivity, beyond a certain concentration more testosterone may have little additional effect. Therefore boys as a group have higher prenatal testosterone and a higher tendency towards hyperactivity than girls, but no association between finger ratio and hyperactivity. Among girls there are low levels of hyperactivity, but a male-type finger ratio is associated with an increase in hyperactivity.

If adults share a common pattern with pre-school children we would expect to find that finger ratios are predictive of personality and behaviour in women rather than in men. Here we have more information, as there have been five studies which support this general conclusion.

First I wish to consider personality. This is a difficult subject because there is no clear-cut consensus among psychologists on how to break down human personality into its component parts. One influential attempt has been to define the following 'Big Five' personality factors or 'dimensions', which are said to represent a complete description of the essence of an individual's personality:[6,7]

(i)   extroversion – extroverts love social situations and show few inhibitions in a group, are often energetic and seek out the company of others; those who score poorly for this dimension are quiet and reserved

(ii) conscientiousness – the conscientious person is well organised and pursues goals with dogged persistence, and to this end is dutiful and methodical; people with low scores in this factor are poorly focused, less careful and easily distracted

(iii) neuroticism – people with a tendency towards neuroticism are prone to feelings of insecurity and emotional distress and often experience negative thoughts and feelings; conversely, low neuroticism scores are found in those who are less prone to distress, more relaxed and less emotional

(iv) agreeableness – cooperative people who score high for this dimension are trusting and cooperative; low scores are reported by people who are often aggressive and less cooperative with those around them

(v) openness – the 'open-minded' person is interested in culture and tends to be imaginative, creative and to be fascinated by educational experiences; those who are more down-to-earth, practical and less interested in art and culture score poorly on this personality factor

These are not value judgements. It is of no interest to the scientist whether it is 'better' to be an extrovert or an introvert, insecure or relaxed, cooperative or uncooperative. What is of interest here is the origin of these aspects of our character. Bernhard Fink of Vienna University, Nick Neave of Northumbria University and I have addressed this question using finger ratios.[8] We took a sample of fifty women and thirty men from Austria and the north-east of England. As expected, the men had lower finger ratios than the women, with averages of 0.96 and 0.98 respectively. However, our main concern was the relationships between finger ratios and the 'Big Five'. As with the pre-school study, females showed the strongest associations. Compared to those with very female-type finger ratios, the women with 'Casanova-type' finger ratios scored low on neuroticism and high on agreeableness. This suggests that the action of high testosterone/low oestrogen on the female foetus influences her adult personality so that she is more likely to be relaxed, less emotional, more trusting and more cooperative. Our results also suggested that such women scored highly on extroversion. However, in this sample the finger ratios had little to say about tendencies towards conscientiousness and openness.

Finger ratios may be more predictive of behaviour in women than men. Can we extract further information on exactly which kinds of behaviour they are associated with? One important aspect of personality is how we respond to boredom. It is tempting to say that a 'Casanova-type' finger pattern may predict an impulsive and adventure-seeking personality, and a study I carried out with Elizabeth Austin of the University of Edinburgh has provided evidence of just such a link between female fingers and behaviour.[9] The study consisted of samples of undergraduates from Edinburgh and Liverpool. The Edinburgh sample was made up of seventy-nine males and eighty-six females, and surprisingly the average finger ratio of this group was 0.97 for both sexes. In the Liverpool sample there were forty-nine males and fifty-one females, with a lower finger ratio in the former (0.97) than in the latter (0.99). As with the 'Big Five' study, personality was measured in such traits as extroversion and neuroticism. There was evidence that high scores for neuroticism were associated with female-type ratios, particularly among women. In addition, the Edinburgh subjects were scored for 'sensation-seeking' – a personality type which shows few inhibitions, is susceptible to boredom and seeks thrills and adventure. Men generally score higher than women for sensation-seeking, and this was confirmed by the proportion of 'Casanova-type' finger ratios for this personality type. However, the associations were again stronger for women than men. Hence a woman with a masculine-type finger ratio may well have a tendency to lack inhibitions and actively seek out thrills. Both neuroticism and thrill-seeking are related to sex in that the former is more common in women and the latter in men. What of the relationship between finger ratio and an overall measurement of a 'sex-role' identity?

On hearing the news of a birth people invariably ask, 'Is it a boy or a girl?' The answer is simply dependent on whether the child has a penis or a vagina. At birth we are assigned to a sex and are socialised accordingly. Most of us identify with the sex to which we are assigned; the sex that we identify with is our gender. Others also assign us to a sex, and this is then our gender role. Gender identity and gender role are usually the same as the sex of rearing. However, there are exceptions.

Some characteristics of personality are often thought of as being typically male or female. One example is assertiveness. Glen Wilson of the Institute of Psychiatry at the University of London has investigated the association between finger ratios and assertiveness in women.[10] Women display considerable variation in this per-sonality trait, which Wilson measured in a novel way. He designed a questionnaire entitled 'Changing Women in the 1980s', which asked women to classify themselves as one of the following: 'assertive or competitive', 'fairly average', or 'gentle and feminine'. The participant was then asked to measure the length of the index and ring finger of her left hand. The questionnaire was published in a national British newspaper and attracted 985 responses. Wilson found that those respondents with a 'Casanova-type' finger ratio reported a higher degree of assertiveness than women with a feminine finger ratio.

Assertiveness is only one stereotyped male characteristic. What of more subtle mixes of femininity and masculinity within individuals? Most men lack many of the characteristics considered stereotypically female – affection, sympathy, sensitivity to the needs of others, understanding, compassion, eagerness to soothe hurt feelings, warmth, tenderness, love of children, and so on – but what is more to our point is that many women lack them too. Similarly, stereotypically male attributes – independence, assertiveness, self-reliance, athleticism, and so on – exclude most women but also some men. Furthermore, these lists of male and female personality types are not mutually exclusive: according to circumstance one may be dominant and tender, aggressive and gentle, masculine and feminine.

Those of us who are perfectly balanced in the expression of masculine and feminine behavioural traits are said to show psychological androgyny. However, many show a preponderance of either masculine or feminine traits. One way to measure this is to use the Bem inventory,[11] a questionnaire containing sixty items, twenty of which consist of male attributes, twenty female attributes and twenty gender-neutral or 'filler' items. Participants are asked to score on a seven-point scale how closely each item describes their personality. Let us take the attribute *competitive* as an example. Do you consider that it is never or almost never true that you are competitive? If the answer is yes score one. The scores increase with stronger agreement until a yes to 'Do you consider that it is always or almost always true that you are competitive?' scores seven. The average of the masculine and positive scores is then calculated, and the masculine scores subtracted from the feminine scores. A negative result indicates the subject has a preponderance of masculine attributes, while a positive result suggests a greater proportion of feminine characteristics.

We are now in a position to consider whether our finger ratios predict our Bem scores of overall masculinity or femininity. When Arpad Csatho and his colleagues at the University of Pecs in southern Hungary examined the relationship between digit ratio and Bem scores in a sample of forty-six female undergraduates,[12] a 'Casanova-type' finger ratio was associated with negative, male-type scores in the Bem, while women with female finger ratios had female Bem scores. Women with 'Casanova-type' finger ratios, then, appear to have a masculine sex-role identity, and those with a feminine finger ratio a feminine sex-role identity. We can now extend this to other, more intensely male traits.

The portrayal of male-typical behaviour often involves physical aggression. In non-human animals the breeding season is the time for much male display, posturing, assessment and even fighting. Males will fight for females or for resources that attract females. The fighting is occasionally intense and is usually associated with increased levels of testosterone in the blood. Parallels with humans are perhaps too easily drawn; work by James Dabbs and others has shown that aggressive and criminal behaviour in men is associated with high testosterone,[13–15] while others, such as John Archer of the University of Central Lancashire,[16,17] have shown no relationship or even an association between low testosterone and violence.

I know of no evidence that shows a straightforward association between finger ratio and aggression in boys or men. Women tend to be neglected in studies of physical violence, simply because they show less of it than men. However, in females there is support for a relationship between one aspect of finger ratio and violence. Women with a masculine-type ratio in their right hand but a feminine-type ratio in the left hand tend to be aggressive. What is the explanation for this remarkable association?

The finger ratio of the right hand is often slightly more 'Casanova-like' than that of the left hand. As we have seen, the observation that the right side has some mysterious connection with all things male and the left with femaleness is reflected in other areas of the human anatomy.[18,19] How does this relate to finger ratios and aggression? Subtract your left-hand ratio from your right-hand ratio. If your right hand has a lower, more masculine ratio than your left you will find that R-L for you is negative. Such a difference is associated with a tendency to aggression in girls and women. If your right is more feminine than your left then R-L is positive. This difference is related to low aggression in females. The evidence for these associations comes from studies on Jamaican school-children and Liverpool undergraduate students. The Jamaican work was part of a long-term project led by Robert Trivers of Rutgers University.[20] The children in this study were from a rural area in the south of the island, and were aged from five to eleven years with an average age of eight years. There were forty-eight boys and thirty-nine girls in the sample concerned with aggression. Measures of aggression were taken from routine teacher reports over a period of two years citing incidents such as pinching, pushing, punching, kicking, jabbing with a pencil, etc. As expected, boys were recorded as being more frequently aggressive than girls, with an average of just over six incidents for boys and just under four for girls. The number of reports showed a similar picture, with boys ranging from zero to twenty-seven incidents and girls from zero to eighteen reports. This is unsurprising, but when the R-L differences in the finger ratios were examined it was found that among girls those with a more 'Casanova-type' ratio in the right hand compared with their left had most reports of aggression. Among boys there was no relationship between R-L differences in finger ratio and aggression.

I believe this indicates that early testosterone has left its indelible stamp on the brain and the fingers of girls, resulting in aggressive tendencies and masculine finger ratios in the right hand. Evidence from measures of physical and verbal aggression, anger and hostility in one hundred Liverpool University undergraduates suggests that this relationship continues into adulthood. This study consisted of a sample of fifty-one men and forty-nine women.

Tendencies towards aggression were measured from responses to the 'Buss and Perry Aggression Questionnaire', which consists of twenty-nine items scored by participants on a five-point scale of agreement. As expected, on average men scored higher on total aggression (79.22 points) than women (72.51 points). In women a low masculine ratio in the right hand compared to left was associated with high aggression scores. Again, there was no such association in men.

Thus far I have been concerned with 'normal' variation within the population. There are other behaviour patterns, however, such as those associated with developmental disorders, which may be seen as pathological or harmful. Autism is one such disorder.

Some of our greatest challenges and rewards arise from the fact that we are social animals. What are those around us thinking, and what is their emotional state? How do they feel about our actions? Do they mean what they say or are they deceiving us? Such questions constitute an everyday puzzle that most of us like to feel we get approximately right. In order to solve such questions we each have what psychologists call a 'theory of mind' – we are able to guess at the emotions of others, to put ourselves in their place, to empathise. Children with autism, however, do not possess this faculty: they show 'mindblindness' to the emotions of others, often treat others as objects, avoid eye contact and disregard cultural norms. They are usually unable to predict the actions of those around them and they have little or no understanding of deception.[21] These are formidable handicaps for a member of a social species.

There is little evidence that parenting is important in the development of autism. On the other hand, there are a number of biological factors involved, and here I am most interested in the many indications that early testosterone may increase the probability of autism.[22] There is evidence that testosterone in the womb reduces language ability but favours the development of abilities in music, drawing and perception of shapes. High prenatal testosterone may also be important in influencing career choices, e.g. in mathematics, physics and engineering. All these factors are associated with autism

– and indeed the majority of affected children are male (out-numbering girls by four to one) and many have problems with language. Some children, however, have normal language develop-ment but all the typical social and communication problems of autism. This condition is called Asperger's syndrome, and again boys outnumber girls, in this case by nine to one. Many children with autism are profoundly handicapped, but some show amazing 'islets' of ability in such things as music, drawing, finding shapes within patterns and some specialised forms of mental arithmetic. The fathers and grandfathers of children with autism are more likely to be engineers than a randomly selected sample of men from the population,[23] and students of engineering, physics and mathematics have reported above average numbers of children with autism among their relatives. Taken together, the abilities associated with autism begin to look like an excessive development of male-ness. It is because of this that Simon Baron-Cohen of Cambridge University has suggested that autism results from the development of an 'extreme male brain'.[2,22]

If testosterone causes autism, how does it do it? One possible answer has been suggested by Soo Downe of the Department of Nursing at the University of Central Lancashire. It concerns the hormone oxytocin. In childbirth, artificial oxytocin is used to stimulate the contraction of the uterus, thus speeding up the passage of the child down the birth canal. Women produce oxytocin naturally, and its functions include the control of uterine contrac-tions and the production of milk. However, it also has an important influence on social behaviour, leading to an increased interest in social contact. Oxytocin is in fact a stress hormone, and its pattern of production and effect is different in the sexes.[24] In response to stress women produce more of the hormone than men, and oestrogen amplifies its effect by influencing the oxytocin receptor molecules found in the brain and elsewhere. Men also synthesise oxytocin, but testosterone reduces its effect by down-regulating the receptors for the hormone. The result of this is that while women tend to respond to stress by seeking company, men tend to remain withdrawn, preferring not to discuss the source of the stress with others.[25] How can we relate these patterns to autism?

It is possible that early testosterone may permanently affect the number and activity of oxytocin receptors in the brain. Thus a foetus which has been exposed to very high prenatal testosterone may have few functional oxytocin receptors, causing massive reductions in its social behaviour as a child, an important aspect of autism. Another aspect of autism, repetitive behaviour, may also be related to oxytocin. Animals in zoos and circuses often show such behaviour, which results from stressful enclosures and lack of contact with others of the same species. They show repetitive mannerisms such as swaying and rocking, self-directed touching, scratching and hair-pulling, and have unusual attachments to objects. This is also typical of many children with autism, and infusion with artificial oxytocin has been shown to reduce these symptoms.[26]

The genes which connect autism and finger ratios may be those that influence testosterone production. If these were the only factors in autism we would expect incidence rates to be stable. However, there is evidence of an increase in autism.[27] In a recent questionnaire addressed to teachers in England and Wales, two-thirds believed that there were more autistic children in primary school than five years ago, and the rate of autism is three times higher in primary than in secondary schools. Similar increases in autism have been reported in Scottish schools (18% up in one year), Cambridgeshire (increase from one in two thousand to one in seventy-five children) and Sunderland (a tenfold increase from 1989 to 1993). However, the major changes have been reported in the US – for example, an increase of 273% in California during the period 1987 to 1998. Undoubtedly some of the increase can be put down to changes in diagnosis and greater awareness of autism, and we need to quantify the effects of these factors.[28,29] However, many feel the changes are too substantial to be entirely explained away in this fashion.

There are at least two environmental factors that may explain an increase in autism. Introduction of the MMR (combined measles, mumps and rubella) vaccine is thought by some to coincide with the increase in autism.[30] However, others have shown that the apparent rise in autism started some ten years before it came into use, and that therefore there is unlikely to be a direct connection.[31] I share the latter view. However, a recent unpublished finger-ratio study (by Peter Bundred and myself) has found that people with masculine finger ratios report a greater incidence of rubella infection than those with a feminine ratio. Rubella during pregnancy can cause autism in the developing foetus; it may be that low-ratio women are less likely to mount a strong antibody response to the rubella component of the MMR. As a result they may be prone to sub-clinical infections of rubella which may then affect the development of their low-ratio foetuses. An association between masculine finger ratio and autism may therefore be found in low-ratio mothers and their affected children.

Then there is the increasing use of artificial oxytocin or pitocin to induce births. In the US this practice has increased twofold since 1990, so that now one in five births is assisted by pitocin; the use of pitocin is more frequent with male births. It may be that pitocin increases the chances of autism in the child, perhaps by causing a sudden increase in testosterone which permanently down-regulates receptors for oxytocin. If this hypothesis is correct, I would expect the finger ratio of women who require pitocin for childbirth to be relatively 'Casanova-like', masculine-ratio women to tend to give birth to babies with masculine finger ratios, and babies with masculine finger ratios to respond strongly to pitocin, i.e. to produce large amounts of testosterone and thus down-regulate their receptors for oxytocin. This could mean that a masculine finger ratio and high pitocin at birth may be found in oxytocin-resistant children. Such a combination may favour autism.

There has been one test of the hypothesis linking early testosterone and 'Casanova-type' finger ratios with autism,[32] comprising seventy-two children with autism of whom twenty-three possessed normal intelligence (i.e. Asperger's syndrome). There was also a sample of relatives comprising thirty-four unaffected brothers and sisters, eighty-eight fathers and eighty-eight mothers. In comparison with individuals drawn from the general population, the children with autism and their relatives had low finger ratios. Furthermore, within the autistic sample those children with Asperger's syndrome had higher ratios than the children with language delay and lowered intelligence. These findings suggest that in families in which autism is found the foetus is exposed to high levels of testosterone, and in a proportion of children this leads to autism.

Autism is the most severe of the childhood psychiatric disorders, but there are others that may result from unusual concentrations of early sex hormones. Attention Deficit Hyperactivity Disorder (ADHD) is one such developmental disorder, which like autism especially affects boys.[33] ADHD usually appears by the age of eighteen months, its symptoms peaking at about three years. It is strongly influenced by genes, defies good parenting and involves one or more of the following: inattention, inability to control impulses, hyperactivity and a low boredom threshold. There is overlap with other, similar conditions, so that 60% of children with ADHD are also oppositional and defiant in their behaviour. The condition is alarmingly common, with about 5% of children and 3% of adults showing symptoms, and 70% of children taking their ADHD into adolescence and 10% into adulthood.

ADHD is so common that many have questioned whether it is a true psychopathology, arguing that it may simply be naughty behaviour. The finger ratio may be useful in understanding ADHD. Like autistic children, those with ADHD are likely to have 'Casanova-type' ratios, but there is considerable variation. One possibility is that testosterone causes low arousal and delayed development in ADHD children, who overcome this through disruptive behaviour. The success of drugs such as Ritalin in treating the condition may come from their stimulant properties. If we assume a testosterone–ADHD link there are a number of questions to be answered in relation to ADHD and finger ratios. What is the average finger ratio for ADHD children? Does finger ratio relate to all aspects of ADHD or mainly to one, such as impulsive behaviour? Does finger ratio predict the effectiveness of drugs such as Ritalin in the treatment of ADHD?

Dyslexia is a specific learning disability[34] whose symptoms are difficulty with writing and spelling, and sometimes with reading and working with numbers. Children with dyslexia find it difficult to 'sound out' words. It is common, occurring in about 8% of the population, and it may be the result of early developmental problems or repeated ear infections. Like autism and ADHD, dyslexia is more common in boys than in girls (three to one), indicating the possible role of high prenatal testosterone. This and the problems with language give us good grounds to suspect the role of testosterone. Dyslexia, then, is likely to be associated with masculine-type finger ratios.

Although there is evidence that children with autism, Asperger's syndrome, ADHD and dyslexia may all have been exposed to high prenatal testosterone, this research is at a very early stage. My

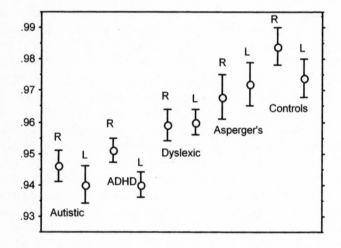

FIGURE 4.1
The average right- and left-hand finger ratios for samples of children with autism, attention deficit hyperactivity disorder (ADHD), dyslexia, Asperger's syndrome and normal children. It can be seen that autistic children have the most masculine finger ratios, indicating very high prenatal testosterone. ADHD children have higher ratios, then dyslexics, Asperger's and finally normal children.

research team has recently measured finger ratios in children with these disorders; Figure 4.1 shows average ratios for the samples. Children with autism turn out to have the most masculine finger ratios, followed by ADHD children, dyslexics, Asperger's and finally normal children. We need more data, but I suspect that this may turn out to be the final position. Clearly prenatal testosterone can have important impacts on the developing brain. Can it also have effects on our predisposition to major diseases?

V

Fingers, Heart Attacks and Cancer of the Breast and Ovary

In developed countries in the twenty-first century we tend to minimise the importance of sex differences. It is suggested that if we argue for substantial differences between men and women we will inevitably make value judgements concerning abilities in one sex compared to the other. In spite of this, I think it is important to acknowledge that influences related to sex reach into almost every aspect of our lives. Sex differences are of course found in organs necessary for reproduction, in adaptations that result in an increase in the survival of offspring (such as milk production by the human breast), in the behaviour and organs which influence the outcome of physical competition for resources that influence mate choice (e.g. an efficient heart and strong muscles), and in those traits which are involved in sexual attraction (e.g. the narrow waist and full hips of women, and the large jaws of men). This all-embracing view of the influence of sex enables us to include such things as brains, hearts and breasts in the search for traits that are associated with male- and female-type finger ratios.

Here I am concerned with important sex-related diseases which occur in adults but whose beginnings lie in the womb. I believe that the pattern and nature of our decline in middle life, and the disease which will eventually lead to our death, is dependent to a large extent on our experiences as a foetus. First I wish to consider heart attack.

As I have mentioned (pp. 16–17), men are more likely to have a heart attack than women. This important sex difference is particularly true of men and women who are young or in early middle age, but it also applies to the population as a whole.[1] For example, in the United Kingdom heart attacks are twice as common in men as in women, and in Europe as a whole the sex ratio of heart attack occurrence is three men to every one woman.[2] However, the sex difference appears to be at its highest in those societies where rates of heart disease are also high. This is the case for the developed world, but in many other nations heart disease is relatively rare, and the sex differences in heart attack rates are real but low. Why is this so?

I recall that in the 1950s it was often said that men have stressful jobs, and stress leads to heart disease. However, the change in sex roles has not brought with it an equal risk of heart attack for the sexes. In the twenty-first century the common explanation for these figures is that oestrogen protects against heart attack but testosterone encourages its early occurrence. This is not even half right. As I have mentioned (p. 25), oestrogen is only a protective against heart attack in women. The relationship between testosterone and heart disease is surprising and counter-intuitive, because it too protects against heart attack, but only in men.[3] This is surprising because testosterone and its derivates the anabolic steroids have long been associated with health risks in athletes and bodybuilders. Changes in the function of the liver, increases in harmful blood lipid levels and in blood pressure have all been reported in otherwise healthy young men who have taken testosterone to enhance athletic performance and muscle gain, and anecdotes concerning steroid-related catastrophic heart attacks in weightlifters and bodybuilders are too common to ignore. The conclusion seems simple – testosterone taken orally or by injection increases the risk of heart disease. The sex hormones are best produced by the body rather than taken orally or injected, so that their secretion is carefully controlled by a complex system of checks and counterchecks. Moreover, it is prenatal testosterone, a hormone produced by the foetus itself, which we are most concerned with here. I believe that foetal testosterone protects males against heart disease and premature heart attack in later life.[4] If I am right then the 'Casanova-type' finger ratio is an indicator of low risk for heart attack. In order to understand why and how this may be so we must take a short digression into the controversial world of sexual selection theory.

Genes survive by being copied and transmitted from generation to generation. A gene that enables men to compete efficiently for resources that attract females is likely to be passed on in the form of many offspring. In societies in which men are allowed many wives the most competitive males can be phenomenally successful in this game, leaving other men without sexual partners and children. In evolutionary terms, sexual selection forces the latter to die out.

I have mentioned the physical differences between men and women. As these differences cannot be explained away by appeals to socialisation or society's expectations, we must instead look for the foetal origins of these sex differences. My belief is that they originate from the relative amounts of testosterone and oestrogen that bathe the developing foetus. If the foetus is male, high testosterone and low oestrogen bring with them the benefits of an efficient heart and blood vessels, an ability to run fast and for long distances, and strong muscles. Indeed, those women who inadvertently take oestrogen-like substances while in the early stages of pregnancy have baby boys with elevated rates of heart and blood-vessel defects and developmental problems of the fingers.[5,6] The evidence also points to testosterone, particularly prenatal testosterone, as a protective against the deterioration of the male heart with age. Testosterone does this because it ensures that the heart and blood vessels are soundly built, without small defects that would hinder the smooth flow of blood to the muscles and the brain. Testosterone does not precipitate heart attacks in men; it prevents them.

Why, then, is heart disease the major killer of men in developed countries? As I have mentioned (pp. 16–26), lifestyle is only part of the story. I believe that in many developed nations a prenatal environment of low testosterone and high oestrogen is commonplace for the developing male foetus. For this reason, in such populations male hearts are not often fashioned with precision, their development is compromised and the consequence is a plague of heart disease, heart attack and premature male death.

Let us consider age, or more precisely age at first heart attack.[7] Through the unemotional eyes of the evolutionary biologist a life-threatening incident that occurs in old people can have little effect on their lifetime family size or 'direct fitness'. This is because they are likely to have had their children, and their effective investment in their family is also drawing to an end. Thus a gene which increases the probability of heart attack in one's seventies or eighties is not going to be strongly removed from the population by natural selection – the frequency of the gene may just drift slowly upwards or downwards. However, if such a gene also confers some advantage in our earlier years it will probably become very common. An accumulation of these kinds of genes is the usual evolutionary explanation for our pattern of vigorous youth and halting old age. However, in Western societies heart attacks, some fatal, are occurring among many men in their thirties and forties. Is this apparent pattern of accelerated deterioration in the male cardiovascular system associated with biological factors, particularly genes, and if so why is it not halted and reversed by natural selection? We can use the finger ratio to consider these important questions.

Figure 5.1 shows the relationship between finger ratio and age at first heart attack in 292 white Caucasian men from the north-west of England. The line through the scatter of points represents an average of these data. Men with a 'Casanova-type' finger ratio tend to have their first heart attack late in life, while those with a female-type finger ratio tend to experience it early. We should note that these observations are of men who have had a heart attack and survived. We do not know if finger ratio predicts age at fatal heart attack. However, it is likely that it does, and it may even be a stronger predictor than for non-fatal attacks.

Although the probability of heart attack is raised in people who are overweight, have a thick waist relative to their hips, smoke and are in a low socio-economic group, statistical tests suggest that finger ratios predict age at heart attack independently of these life-style factors. However, these same tests indicate that lifestyle factors remain independent indicators of raised risk of heart disease.

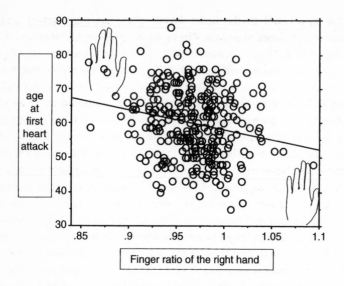

FIGURE 5.1
Finger ratios and age at first heart attack in 292 white Caucasian men from
the north-west of England. In this sample heart attacks occurred as early
as the mid-thirties to as late as the mid-eighties. Men with a female-type
finger ratio were more likely to experience their heart attack when in their
thirties to early fifties. Subjects with a 'Casanova-type' finger ratio tended
to have their heart attack in their mid-fifties to eighties.

It will take some time to disentangle just what predicts what.
However, the latest findings indicate female-type finger ratios in
men are related to high levels of fibrinogen and thyroid stimulating
hormone (TSH). These associations may hold the key to the power
of finger ratios as indicators of a healthy heart.

Fibrinogen is considered by many as a key factor in explaining
heart disease.[8–10] It is found in the blood in large amounts and
functions as an important part of the clotting process. Suffer a cut
and a cascade of molecular effects occurs in the damaged blood
vessels; eventually a clot made up primarily of fibrinogen is formed,
preventing further bleeding. The possibility exists that this process
may go awry, leading to the inappropriate formation of a blood clot

that may travel to the heart, blocking the coronary artery and starving the heart of oxygen. The result is a myocardial infarction or heart attack. In men, female-type finger ratios are associated with high concentrations of fibrinogen, but we do not know whether they are also connected with a tendency to spontaneous clotting. There is more to fibrinogen, however, than a propensity to fatal clotting. It is a very large molecule, and there appears to be much more of it in our blood than we need to maintain the integrity of our vascular system. The result is that the hearts of men with high fibrinogen levels must labour to pump their viscous blood around their body (this might provide the link between running speed and masculinised finger ratios). Men with 'Casanova-type' ratios have low fibrinogen and low-viscosity blood which slips easily through narrow capillaries. High blood viscosity may result in continually high resistance to the action of the heart, which could wear it down. This long process of attrition may then accelerate the age-related process of vascular decline. Again, lifestyle factors could be relevant. For example, weight gain, increase in stress, smoking and ageing have all been linked to increasing fibrinogen concentrations.

Although a major factor, fibrinogen cannot be the only variable in the decline of our heart and blood vessels. A recent as yet unpublished study by Peter Bundred, Charles van Heynigen and myself has indicated that finger ratios correlate with fibrinogen, while weight and waist-to-hip ratio are powerful indicators of lipid levels. High cholesterol in our blood stream, particularly of the low-density variety, is an established indicator of risk for serious heart disease. Lower your low-density lipid levels and your chances of having a heart attack reduce accordingly. This part of your risk for heart attack is not strongly indicated by finger ratios. Thus I conclude that there are very simple body measurements which are independently associated with the efficiency of the cardiovascular system. Weight and waist-to-hip ratio are well-known and useful measures which predict different aspects of risk than finger ratios. However, their use is not straightforward: they are not fixed, and we are forced to adjust them according to our age. Finger ratios can function as a very early warning sign of a poorly formed vascular system and high levels of fibrinogen. Can they point to even more?

A precision-made vascular system and a healthy blood viscosity are not the only components of a dependable heart. Arrhythmia or irregularities of the heartbeat may be a further risk factor for heart attack. Such irregularities may arise from a malfunction of the nervous system or from an abnormally high metabolism. Finger ratios may predict some aspects of metabolic level through their relationship with the thyroid gland.

The thyroid consists of two lobes found on each side of the windpipe or trachea, connected by a thin strip of tissue. It is larger in women than men, and enlarges during menstruation. The principal products of the gland are the iodine-containing hormones thyroxine (T4) and the more active tri-iodothyronine (T3), which is produced from T4. Synthesis of these hormones starts around the third month of foetal life, and shows increases at puberty and pregnancy. However, it is in the pattern of thyroid disease that we see the most intense sex differences. Women are more prone to thyroid disease than men. In adults T4 deficiency leads to puffiness of the face, scanty eyebrows and enlarged blood vessels in the cheeks. This syndrome is called myxoedema and is most common in middle-aged women. Some geographical areas are deficient in iodine, resulting in cases of massive enlargement of the thyroid. This is goitre, and again it is more common in women.

How does the thyroid gland affect the body? Remove the thyroid and body temperature falls, pulse and respiration slow, mental apathy settles in and the patient gains weight. Enlargement or hypertrophy of the thyroid gland leads to the production of too much T4 and T3. The heart races to 150 to 160 beats per minute, thought processes and activity levels race along with it, fat stores are broken down and exhaustion is common – unsurprisingly, this condition leads to symptoms of heart disease,[11] and indeed Peters and his colleagues[12] have found that high levels of T4 are associated with a threefold increase in the chance of heart attack. The rate of T4 production is controlled by the pituitary, a gland which lies within the head, closely associated with a part of the brain called the hypothalamus. Signals from the hypothalamus and the blood supplying the pituitary lead to the production of thyroid-stimulating hormone (TSH). High levels of TSH stimulate the thyroid gland to produce T4, increasing metabolic and heart rates. An increase in T4 in the blood leads to a reduction in TSH production: T4 reduces, and heart rate goes down. If metabolic control is set so that TSH is consistently produced in high amounts this could lead to an elevated heart rate, arrhythmias and even heart attack.

A recent study of children in the North-West Province of China[13] suggests that female-type finger ratios are associated with high levels of TSH in the blood. Iodine levels tend to be rather low in the soils of this area. Following a programme of iodine supplementation, photocopies were taken of the hands of 140 children. TSH and T3 levels in these children were measured from blood samples. In order to establish the patterns of finger ratios in the area, photocopies of the hands of an additional sample of 417 children were taken. As mentioned earlier (see pp. 29–30), there are two main ethnic groups in the North-West Province, the Uygurs and the Han. It turned out that Uygur children had more masculinised finger ratios than Han children. With regard to the levels of TSH and T4 there were ethnic and sex differences apparent within the samples. After controlling for these there emerged a clear trend. Children with female-type finger ratios had higher concentrations of TSH than children with 'Casanova-type' ratios. There was also a weaker relationship between high finger ratio and elevated T4 levels. If these trends can be found in other ethnic groups it could mean that high finger ratios in men predict chronic patterns of thyroid stimulation, consequent high T4 levels, elevated heart rates and arrythmias.

Before we leave the subject of finger ratios and the vascular system we should consider blood clots that initially occur away from the vicinity of the heart. Thromboses are not always formed as a response to an injury. They may occur spontaneously, and occasionally they can move in the body, blocking important blood vessels and leading to catastrophic strokes or heart attacks.

There are a number of factors that may make us more likely to form such a blood clot. There is much debate at present over whether long-haul jet flights, with their restriction in leg room, enforced lack of exercise, tendency to cause dehydration and to increase blood viscosity, may provide an environment which increases the likelihood of deep-vein thromboses. I would suggest that passengers, especially male passengers, with female-type finger ratios are particularly at risk of thromboses that arise as a consequence of an increase in blood viscosity. It may be worth using finger ratio to indicate passengers at greatest risk, so that encouragement to exercise and to drink water (and not dehydrating alcohol) can be targeted most efficiently.

In women breast cancer exacts a toll similar to that of heart disease in men. Although breasts develop in female and male foetuses in essentially the same way, and breast tissue is present in men as well as women, almost all sufferers of breast cancer are women, with cases of male breast cancer making up less than 1% of the total mortality rate.[14]

In attempting to explain this sex difference, it is worth noting that incidence of the disease is not uniform across geographical areas. Breast cancer in the developed world accounts for 17% of all cancer deaths for women, while the rates for developing countries are substantially less, at 12%. This pattern is complicated, however, by the fact that in developed nations such as the US and the UK incidence of breast cancer in white women is five times higher than in oriental women in developing China and affluent Japan.[15,16] This difference may to an extent be genetic, but there is evidence for a strong environmental effect. Migration to the West is associated with increased levels of breast cancer in second-generation east-Asian women.

Oestrogen is the key to understanding breast cancer.[17] The breast increases in size during puberty and begins to mature, a process that is associated with rapid cell division. Division is most marked in the epithelial cells lining the complex of ducts within the breast. Oestrogen stimulates division of epithelial cells, especially after puberty and before first pregnancy. Cell division can lead to errors in copying of genes, while the breakdown products of oestrogen may directly damage genes. Breast cancer may be triggered by genetic lesions.[18] However, the influence of oestrogen does not stop there. Many breast tumours are dependent on a plentiful supply of the hormone for their growth. Numerous studies have found receptors for oestrogen in 50% to 80% of breast tumours. Block the attachment of oestrogen to the receptors with such drugs as Tamoxifen and you may slow the growth and spread of the tumour. More than half of the tumours that are rich in oestrogen receptors respond to oestrogen-blocking drugs and other hormonal treatments, while those without oestrogen receptors respond to such treatment in only about 5% of cases.[19] Our knowledge of oestrogen and breast cancer can be used to treat established tumours. Can it also be used to identify high-risk groups who may be screened early?

There are a number of risk factors for breast cancer, and all point to the importance of oestrogen in its origin. An early start to periods, late first birth, few or no children, late menopause – all are associated with an increased chance of breast cancer and an increased lifetime exposure to oestrogen. The reason can be readily seen from our knowledge of a typical cycle. At the beginning of the cycle oestrogen levels are low. However, as the cycle progresses and the egg matures in the ovary, oestrogen levels rise. The peak is reached just prior to ovulation, when the hormone level may be many times that found at the start of the cycle. Hence a woman with many ovulations will also experience many oestrogen surges.

Cancer rates are highest in Westernised populations. Boyd Eaton and his colleagues[18] have estimated the expected number of ovulations experienced by women in developed countries. In such countries as the US and the UK the age at first period or menarche has been getting steadily younger. In the US it is now approximately 12.5 years. The average age at menopause is 50.5 years and, ominously, this may be rising. This gives us thirty-eight years of ovulations. Multiply this by thirteen ovulations per year and we have a possible 494 surges in oestrogen. However, we must deduct from this the effect of carrying children and the suppression of ovulation by breastfeeding. In the US women have on average 1.8 children and breastfeed for about three months. This means that as a result of reproduction they suppress about twenty-three ovulations. Subtract a figure of twenty for anovulatory cycles, miscarriages and stillbirths, and what remains is 450 potentially cancer-causing surges in oestrogen.

It is questionable that women are adapted to experience such a blizzard of 'oestrogen-hits' during their reproductive life. In the not-too-distant past our ancestors lived a hunter-gatherer existence. There are a few hunter-gatherer groups left, and in these populations women do not experience such a high oestrogen load. Their average age at menarche is later (16.1 years), and age at menopause earlier (forty-seven years). This gives 30.9 years in which they ovulate, and a possible number of oestrogen surges of about 402. However, their completed family size of about six children is higher than that of their Western counterparts, and a further substantial reduction in ovulations occurs through an average breastfeeding period of about two years. As a result, roughly 215 ovulations are suppressed. If we include anovulatory cycles, stillbirths and miscarriages, a hunter-gatherer woman may expect to experience about 160 ovulations. This means that women in developed countries have about three times the oestrogen exposure of their recent ancestors.

The birth-control pill modifies the pattern of oestrogen exposure, but it is not possible to quantify an exact increase in cancer risk. Delaying first pregnancies, however, unquestionably makes women more prone to breast cancer. This lifestyle pattern is characteristic of Western women and particularly those educated beyond the age of eighteen years. The risk factors for breast cancer among women in non-developed nations lie somewhere between those of developed nations and hunter-gatherer societies. It is not surprising that breast cancer is most common in the West.

Although the overall pattern of high oestrogen exposure in Western women goes some way to explaining why developed nations show high rates of breast cancer, there is a real concern that many cases of breast cancer are left unexplained. Dimitrious Trichopoulos may have identified another factor indicating increased susceptibility to breast cancer.[20] He has pointed out that known risk factors for breast cancer do not fully explain the major differences in breast cancer rates between countries. The missing factor, he argues, is oestrogen, but oestrogen applied to the foetal breast and not the adult breast. Oestrogen levels during pregnancy are ten or more times higher than at other times, and these concentrations vary substantially from woman to woman. This is exactly what we would expect if breast cancer can originate in the womb.

We now have a working hypothesis for the missing risk factor. Let us compare two patients with breast cancer. In one the tumour became obvious when she was aged thirty-five, while in the other the tumour was discovered when she was seventy. The difference in their age at diagnosis may reflect their oestrogen exposure when they were in the womb. High oestrogen at this time may have damaged the genes within the developing breast tissue, eventually leading to breast cancer. It can be expected that other genes may modify and delay the formation of the tumour, because the longer the delay the greater the likelihood of modifying genes being passed on to the subject's children. However, even these genes may not be able to suppress the inevitable when genetic damage is extensive. The result is a tumour relatively early in life. When levels of oestrogen in the womb are low, damage to the genes of the developing breast may also be low. Modifying genes are able to suppress breast cancer. However, such genes are likely to be ineffective in old people, because their operation at this point will not increase their chances of being passed on. Cancer therefore occurs.

We can test this simple theory by using the finger ratio, for it bears testament to oestrogen levels in the womb. If we are right, early breast cancer should be related to long index fingers and short ring fingers, while women with late breast cancer should have short index and long ring fingers. There has been one test of this theory in a sample of just over one hundred women with breast cancer.[21] Figure 5.2 indicates that in this sample there is indeed a relationship between female-type finger ratios and early breast cancer. However, one small sample does not prove the argument.

There are other ways of testing Trichopoulos's theory of foetal oestrogen and breast cancer. Suppose the differences in rates of breast cancer between countries were mainly dependent on differences in

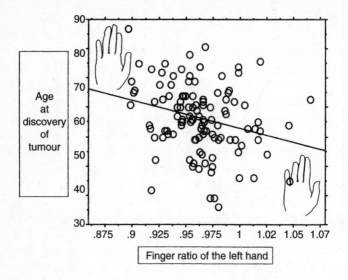

FIGURE 5.2
Long index fingers and breast cancer. This sample of 118 women shows early occurrence of breast cancer in patients with female-type finger ratios.

foetal oestrogen. Is it the case that on average a foetus in the UK experiences five times the oestrogen of a foetus in China? Can we take this further and include other cancers which may be caused by oestrogen? Ovarian cancer is an obvious choice. The ovary is a major source of oestrogen in pre-menopausal women. Boyd Eaton and his colleagues have shown that the risk factors for breast cancer of early menarche, small families and late menopause are also risk factors for ovarian cancer.[18] We now have samples of female finger measurements from eleven populations, representing northern, eastern, southern and western Europe, the Caribbean, southern Africa, South-Central Asia and South-East Asia. These countries show very marked differences in breast cancer mortality rates. Figure 5.3 shows that a steady rise from a 'Casanova-type' finger ratio to more female-type fingers closely corresponds to an increase in mortality from breast cancers. A similar and even stronger relationship applies for ovarian cancer (Figure 5.4).

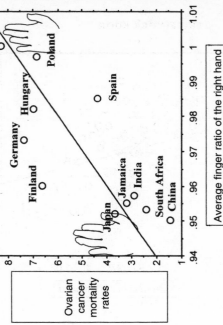

FIGURE 5.3
The finger ratio and breast cancer mortality rates. Death rates from breast cancer vary dramatically between countries. Women in countries such as the UK, Spain and Poland have long index fingers in relation to their ring fingers and high rates of breast cancer. Short index fingers and low rates of breast cancer deaths are common in women in China, Japan and India.

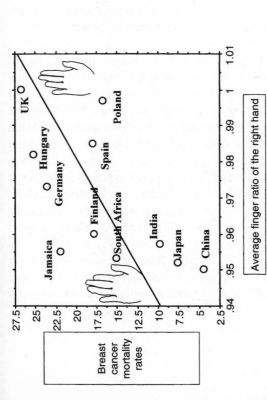

FIGURE 5.4
Finger ratio and mortality from ovarian cancer in eleven countries. Populations with female-type ratios with long index fingers have high death rates from ovarian cancers.

I would conclude that sex differences in disease do matter, particularly when the organs concerned are subject to strong sexual selection (e.g. breasts, ovaries and hearts). This may seem obvious, but there are other, more subtle, differences in how men and women respond to disease, which may also be the result of our experiences in the womb.

VI

Fingers and Infectious Diseases

Why are infectious diseases more frequent in some countries than in others? An obvious response is that countries with a lot of disease are hot and wet – ideal conditions for bacteria to thrive – and that many have overpopulated areas with poor sewage treatment and dirty water supplies, encouraging infection. These and other conditions are indeed related to the prevalence of infectious disease, but they are not the only predictors of disease load.

In general, developed countries do not suffer from a heavy burden of infectious diseases. However, the disease load has shifted to such things as heart attack, breast cancer and atopic diseases – disorders of the immune system such as hay fever, eczema and asthma.[1] Atopic diseases result when the immune system is over-active in its production of the allergy antibody IgE. Antibodies are one response of the immune system to substances or antigens that are found on the surface of invading disease organisms or on the surface of such things as pollen grains and mould spores. IgE, the fifth class of antibody to be discovered (hence the letter E), activates the inflammation so characteristic of atopic diseases. It does this by stimulating the release of histamine from special cells called mast cells. It is the histamine that causes the inflammation. Atopic reactions in the nose (hay fever), lungs (asthma) and on the skin (eczema) are most common in children. For example, in the US atopic eczema affects about 10% of infants but only about 3% of adults.

If atopic diseases are most common in countries with few infectious diseases, it is perhaps the very lack of such pathogens which produces an environment that encourages allergy. This is the focus of an influential theory concerning the origin of asthma, which suggests that a clean, healthy environment with few pathogenic viruses, bacteria and parasites may have an effect on a child's maturing immune system. In such conditions there may be a breakdown in the child's ability to recognise self from pathogens. As a result the immune system turns in on itself, leading to asthma and other disordered immune responses such as hay fever. This, then, is the 'hygiene hypothesis'.[2]

I would like to suggest a new hypothesis that concerns finger ratios and their relation to infectious and atopic diseases. I do not think that we can convincingly explain the distribution of infectious diseases by vague appeals to hot, wet climatic conditions or even poor sewage and water systems. Nor do I think that the 'hygiene hypothesis' adequately explains why populations with few infectious illnesses have high rates of atopic disease. Instead I would propose that populations in which high testosterone before birth is common have, as a consequence, reduced immune function. This leads to a heavy load of infectious disease and very high infant and adult mortality. This effect of testosterone on the immune system does not include acquired immunity to common diseases in a particular area, nor does it relate to single genes which have become common because they give resistance to a specific disease – the association is between testosterone and a general reduction in the level of immune function. However, there are some advantages to reduced immunity, the most important being a reduction in immune attacks directed at one's own body. We should not be surprised that asthma and related diseases are common where infectious disease is rare. In such populations immune function is not reduced by high prenatal testosterone.

Pathogens or harmful disease-causing organisms include many bacteria, viruses, unicells and parasitic worms. Such things exact a dreadful toll of poor human health and mortality. The key to understanding the distribution of these organisms is sex or, more precisely, marriage patterns.

The strength of human sexual selection in any one group

depends largely on the marriage pattern adopted by that group. The pattern varies from polyandry (more than one husband at a time), through monogamy (one wife/husband at a time) to polygyny (more than one wife at a time).[3] Polyandry is very rare and in most cases, e.g. the Lepcha of northern India and in Tibetan society, involves brothers sharing one wife. Monogamy and polygyny, on the other hand, are common, and exert profound influences on sexual selection in many human societies. I will therefore consider their potential effects on our immune system in some detail.

Many societies allow or encourage polygyny, meaning that competition for wives is intense. In such a situation anything which increases a man's ability to compete with other men will be favoured and is likely to be passed to his sons. Thus polygyny will encourage the evolution of high testosterone production in the foetus, as this ensures strength and high cardiovascular function in males. However, testosterone reduces the efficiency of the immune system, and we might expect a relationship between polygyny and infectious disease, among women and men. Prenatal testosterone in girls is reduced by the action of genetic modifiers, but such modifiers do not fully suppress the expression of genes for high testosterone. Hence in polygynous groups infectious diseases should be common in both sexes.

Bobbi Low of the University of Michigan has shown that there is a strong association between disease and polygyny.[4] Polygynous groups have more pathogen stress than monogamous groups. Furthermore, the intensity of competition between males within different forms of polygyny is related to pathogen stress. For example, in 'sororal' polygny a man may marry more than one woman, but they have to be sisters. This form of polygny is not closely related to intense male competition or to the high incidence of disease-causing organisms. However, where men compete physically for wives outside the family group, especially by capture of women from other groups, then we find high pathogen stress. Such competition places emphasis on strength, speed and accurate visuo-spatial judgement for fighting. I would expect such groups to have high prenatal testosterone and lowered immune competence, leading to high pathogen loads. Strictly enforced monogamy relaxes the pressure for high prenatal testosterone, as it is associated with low male competition for wives. As a result, monogamy will be related to low prenatal testosterone, high immune function and low frequencies of infectious diseases.

The usual explanation for the polygyny–pathogen association is that polygyny is adopted as a response to pathogen prevalence. Having many wives, the argument goes, increases the diversity of a man's children, ensuring that at least some of them are resistant to common diseases. However, in polygynous groups many men are unable to reproduce, and so polygyny reduces the effective size of the male population, bringing about a marked reduction in the variability found in children and indeed in the population as a whole. The hypothesis that polygyny is a cause of pathogen load, not a response to high frequencies of disease-causing organisms, explains much about disease, sex and ethnic differences. Let us see how finger ratios help us verify our speculations.

I wish to begin by discussing the spread of a new pathogen, the HIV virus, in relation to what is known about the geographical distribution of finger ratios. HIV is the causative agent of Acquired Immune Deficiency Syndrome (AIDS). HIV/AIDS came to the notice of Western medicine as a result of its transmission between homosexual men and its occurrence among intravenous drug-users. The virus spread quickly – in 1998 more than 20 million people died of AIDS, and by 1999 it was the world's leading killer infectious disease. However, the infection is no longer a Western problem confined largely to gay men and drug-users who share needles. More than 67% of people with HIV/AIDS are now found in sub-Saharan Africa.[5,6] This is a very special area with regard to prevailing marriage systems. In the world as a whole monogamy is found in about 20% of all societies, with occasional and intense polygyny making up 50% and 30% respectively. In sub-Saharan Africa monogamy is present in only 5% of peoples and polygyny in 80%, of which intense polygyny is found in 78%. Therefore south of the Sahara intense male competition for women is common among indigenous peoples.

AIDS is having an appalling effect on the health and life expectancy of black Africans. Within southern Africa life expectancy has been cut by a quarter, and AIDS-related infections such as tuberculosis are increasing rapidly.[7] In sub-Saharan Africa HIV is transmitted predominantly by sex between men and women and vertical transmission from mother to baby. The expected rapid spread of HIV in Western nations, from homosexual or bisexual men to heterosexuals, has not happened. However, it has spread rapidly through heterosexual black populations. I believe this situation has arisen because prenatal testosterone reduces the effectiveness of immune responses to HIV.

Gay sex is associated with risk factors that are likely to increase transmission rates of HIV. For example, anal intercourse may often result in tearing and bleeding, leading to infection. Promiscuity has also been reported to be high among some groups of gay men. However, there is evidence that early testosterone may also play its part in gay susceptibility to HIV. For example, the formation of the penis is dependent on testosterone, and, as I have mentioned (p. 10), homosexual men may on average have longer penises than straight men.[8] Similarly, left-hand preference may be associated with high prenatal testosterone, and gay men show higher rates of left-handedness than heterosexual men.[9] Therefore there is some evidence that gay men are more testosteronised than straight men. As we shall see (pp. 121–9), the situation is less clear when we look at finger ratios in gay men, but homosexuality in men may, on the whole, be associated with high prenatal testosterone.

With regard to sub-Saharan Africa, the available evidence from finger-ratio studies points strongly to very masculinised ratios in black South Africans. For example, among Zulu men in the area of Durban an average ratio of 0.94 is found.[10] Contrast that to the averages of 0.98 to 1.00 found in England, Poland and Spain. Western populations may represent infertile ground for the heterosexual spread of AIDS simply because they are not as strongly masculinised before birth as black populations. In contrast to Europe, the black population of South Africa is suffering from a disastrous HIV epidemic and has 10% of all the world's HIV/AIDS cases. Masculinised finger ratios may therefore indicate susceptibility to HIV infection and perhaps to the rapid development of AIDS once infection has taken place. The problem may be compounded by transmission from mother to child. If a woman's partner has HIV and a masculinised finger ratio he is likely to father a child who also has a masculinised ratio. The child may then be particularly susceptible to infection from the mother.

Before leaving this subject I will mention skin colour. Sub-Saharan Africa has some of the blackest indigenous peoples in the world. In the next chapter I will present evidence that skin pigment is a protective response to immune suppression caused by prenatal testosterone.[11] Thus melanin may block the entry of bacteria which cause genital sores which themselves provide entry points for the HIV virus. Within sub-Saharan Africa the groups with the darkest skin have the lowest HIV rates.[12] I believe this offers further evidence of an association between prenatal testosterone and HIV susceptibility.

There are other explanations for the present distribution of HIV/AIDS. For example, a gene (CCR5) confers some resistance to the spread of HIV within the body.[13] HIV destroys the immune system by attaching to and entering the white blood cells or T cells that are an important part of our immune response. The virus does this by recognising and attaching to the CCR5 protein that is present on the surface of the T cells. However, CCR5 can vary from person to person. HIV has difficulty recognising and binding to a variant protein called CCR5⁻. Therefore individuals with the CCR5⁻ protein are less susceptible to the virus. In northern Europe about 8% of the population have CCR5⁻; in southern Africa it is virtually absent from black populations. However, interesting though this protein is, the differences in frequency are not sufficient to explain the large discrepancy between European and African rates of infection.

Viruses take some time to spread, so that the site of origin of HIV may explain some of the geography of infection. The available evidence suggests that HIV originated in Africa, and this may explain why the virus is now centred on Africa. HIV-1, the most common form of the virus, may have jumped from chimpanzees to humans.

Similarly, HIV-2 may have its origin in a monkey, the sooty mangabey. Eating infected 'bush' meat is one likely initial source of the virus. The reason we are surprised that most cases of HIV are now found in Africa is that there has been ample opportunity in Europe for HIV to move from the gay population to heterosexuals, yet this large-scale spread has not happened. This may be because in general Europeans experience low testosterone before birth.[14]

There is yet another explanation for the introduction of HIV to humans. Edward Hooper has controversially suggested that HIV originated from human polio vaccines that were cultured in primate cells.[15] The batches of the vaccine he implicates were used in two large-scale trials – one in Poland and the other in Africa. Hooper is unable to explain why the Polish vaccine did not cause an HIV epidemic but the African vaccine did. Poland has one of the lowest HIV rates in the world; it also has one of the highest finger ratios. We might expect that the HIV virus would not find the Polish population to be susceptible to infection, but that African transmission would be rapid. There is no convincing evidence that primate cells were used to culture polio vaccine or that HIV originated from such cultures. However, there is an intriguing correspondence between Hooper's speculations and the predictions of a hypothesis for the foetal origins of susceptibility to HIV.[14]

Now let us turn to a consideration of the link between males and disease. A study of the geography of new pathogens and finger ratios may give us clues to the effect of prenatal testosterone on the immune system. Similar clues may be available to us when we consider infection rates in males and females. It is well known that males are usually more susceptible than females to infection by parasites. The clue to understanding this appears to be testosterone. Many vertebrate species such as birds and mammals show marked sex differences in parasite loads, and almost invariably it is males who suffer the greater load. Vertebrates produce testosterone, but invertebrates such as insects do not, and accordingly we do not find a consistent male bias for high parasite loads among them.[16]

As a species humans follow the general rule for infection patterns in vertebrates. Men are less competent to deal with infection than women. For example, one important aspect of immune function is the production of antibodies against micro-organisms. Men have lower concentrations of antibodies in the blood, the connective tissue and on the mucosal surfaces in the nose, sinuses, lungs and gut.[17,18] Furthermore, antibody responses to infectious agents such as polio, measles and rubella are often weaker in men compared to women.[19-21]

Many tropical diseases are more prevalent in males – river blindness, elephantiasis, Chaga's disease, sleeping sickness, kala azar and blood disorders such as Burkitt's lymphoma. Not all of these will be familiar names, but all show aspects of the host response which are less favourable in males than in females.[22] Take, for example, the parasitic worm *Onchocerca volvulus*. This nematode or roundworm is common in sub-Saharan Africa and in Central and South America, and is spread by the blackfly or buffalo fly. Once inside its human host a female worm may live for nine to eleven years and produce up to three thousand tiny microfilariae per day. High worm burdens result in itching, rapid ageing and blindness by the host's mid-thirties. This 'river blindness' is frequent in fertile river valleys, preventing their effective cultivation. It afflicts approximately 1 million people worldwide, and in Africa forty thousand new cases occur each year.[23] As with many tropical diseases, men are more likely to suffer from high loads of the parasite than women. It may be expected that both men and women with masculinised finger ratios will be at particular risk of infection by this parasite.

Ethnic and sex differences in disease susceptibility can give us clues as to how finger ratios may relate to immune function. However, for more direct evidence we must look at fingers and patterns of infection.

To date there has been one study that addresses these issues. The subjects were one hundred young men and one hundred young women undergraduates attending the University of Liverpool. The participants had their right- and left-hand finger ratios measured and all answered a questionnaire that related to their history of illness. The questionnaire was divided into three parts. First there were questions relating to recent illnesses – absence from university through sickness, frequency and duration of colds, and so on – then questions relating to common infectious diseases such as athlete's foot, dandruff, measles, mumps, whooping cough, rubella, chicken-pox and glandular fever, and finally questions on atopic diseases – allergies, asthma, hay fever and eczema. In many cases the responses of the subjects were related to their sex and to their finger ratios, particularly their right-hand ratios.

Overall, participants with low ratios in the right hand reported more illnesses, more colds and longer duration of colds than sub-jects with female-type right-hand ratios. Masculine finger ratios are of course more likely to be found in men, who may or may not complain more about colds and their duration, but these effects are best seen when we look at men only. Here we find the strongest relationships between illness and right-hand finger ratios. Men with masculine ratios reported more illness and more colds that lasted longer than men with feminine-type ratios. These results may reinforce gender prejudices about common illnesses, but they are just as we would expect if early testosterone reduces immune func-tion, is higher in males than females (which it is) and varies in amount from individual to individual.

What about a longer-term view of infectious illness? Measles, mumps and whooping cough are the subjects of mass vaccination campaigns in the UK. Vaccination may hide underlying patterns of susceptibility to disease. Unsurprisingly, it was found that finger ratios did not differ between those who reported having these dis-eases and those who had escaped infection. Rubella may cause birth defects. It is therefore the subject of a vaccination campaign which is directed towards young girls and girls at puberty. Despite this, both men and women with 'Casanova-type' finger ratios in the right hand were more likely to report that they had suffered a bout of 'german measles'. Vaccination against chickenpox is not wide-spread in the UK. However, as with rubella, masculine ratios of the right hand, particularly in men, were associated with infection with this virus. Dandruff and athlete's foot are irritating but less serious than many of the common 'childhood' infections. Both showed an association with low finger ratios of the right hand, and in both the effect was slightly stronger in women than in men. A fairly clear pattern emerges, then: masculine ratios of the right hand indicate susceptibility to rubella, chickenpox, dandruff and athlete's foot. However, there was one interesting exception. There were eight participants who reported a bout of glandular fever. All had feminine ratios of the right hand. When we consider atopic diseases, the trends are reversed. Having a female-type ratio was associated with a susceptibility to allergies (right-hand ratio) and eczema (left-hand ratio) in men, while both men and women with female-type ratios of the right hand were susceptible to asthma and hay fever.

We must beware of placing too much importance on results from one study. Nevertheless, these data point to a general relationship between masculine finger ratios and susceptibility to infectious diseases. Diseases of a disordered immune system which turns in on itself, on the other hand, were associated with female-type finger ratios.

This chapter has presented a new and perhaps controversial view of human disease, but I believe that the theoretical framework I have proposed fits with what we know of patterns of finger ratios and susceptibility to infectious and atopic diseases. In the next chapter I will argue that this hypothesis and our knowledge of finger ratios can also shed light on the evolution of that most contentious of human traits, skin colour.

VII
Fingers and Skin Colour

There have been a number of attempts to divide humans into racial groups. Many classifications recognise five races – Caucasians, East Asians, blacks, Capoids (Hottentots and Bushmen) and Australoids. The definitions of these groups are vague and unreliable, and the genetic diversity within them huge, much greater than the genetic differences between races. Classification of humans into races is therefore of doubtful biological value. Skin colour is often thought of as a marker for these groupings – Caucasians often have 'white' skin, East Asians have yellow pigmentation, and dark skin is common for the remaining races. However, it is not as simple as that. Dark skin is common among populations that experience high intensities of ultraviolet (UV) light, hence Caucasians who have migrated to India and Sri Lanka have evolved heavily pigmented skin, although their facial features and hair indicate that they are Caucasian. Nor are dark-skinned peoples uniformly black. For example, Australoids have dark skin but show a gradation from darkest in the north of Australia to lightest in the south. Therefore skin colour is not bounded by race, but it is under the influence of natural selection.

In this chapter I wish to suggest a hypothesis concerning the distribution of skin colour.[1] UV intensity is indeed an important variable that leads to selection for colour. However, the influence of sex and its correlates, in the form of marriage systems, testosterone levels and pathogen resistance, are also key variables which help to explain why one population is black while others are yellow or white. I will use finger ratios to construct and support my argument, but first we must discuss what is known about the biology of skin colour.

Human pigmentation is dependent on the synthesis of the visible pigment melanin by a special type of cell, the melanocyte. This cell packages the pigment into membranous bags or melanosomes. The pheomelanosome contains light-red or yellowish melanin. Black or brown melanin is contained by the eumelanosome. People with white skin have melanosomes with a highly acidic content which appears to inhibit melanin production. Melanin and the type of one's melanosomes, then, determine skin colour.[2]

Melanocytes are found throughout the body, but are particularly common in the skin, the hair follicles and the eyes. Within the skin they make their melanosomes and export them through fine tubules to the surrounding cells. These take up the melanosomes and become pigmented. The kind of melanocyte one has, and its distribution within the skin, is determined early in development by genes rather than exposure to sunlight. The type of pigment genes present in human populations is controlled by natural selection. In order to understand why indigenous peoples of Europe have evolved to be white, while those of China are yellow and Africans and South Asians are black, we need only identify the selection pressures controlling the frequency of genes for pigmentation. It sounds simple, but this is where we run into difficulties when we consider the function of melanin.

First we need to state the obvious. Melanin absorbs light. The more melanin skin has the darker it is. Skin colour is often measured in terms of its reflectance (the amount of light reflected from its surface) – white skin has high reflectance, black skin has low reflectance.[3] Melanin has a role in protecting the skin from UV light. It is important in maintaining the function of the sweat glands, in preventing the breakdown of important substances in the blood supply to the skin (e.g. folates essential in the development of the nervous system), protecting the production of sperm, and in preventing aggressive cancers or melanomas of the skin. However, melanin reduces the production of vitamin D in the skin. Vitamin D is important in the control of calcium and in bone formation. White skins are therefore favoured when light intensities are low.[4]

Take a journey in your mind's eye from the Equator in Africa to the north of Scandinavia, and the skin of the inhabitants progressively lightens. Nina Jablonski and George Chaplin of the California Academy of Sciences have shown that along this particular Old World line or transect the theory that skin colour is an adaptation to UV radiation seems to hold true.[4] However, there are some exceptions. Why are many African populations, e.g. those of Cameroon, Chad, Liberia and Nigeria, blacker than they should be with regard to UV intensities, and why are the Algerians, Libyans and Tibetans lighter than expected? Jablonski and Chaplin's sample consists of eighty-five populations. Sixteen of these are darker than expected, twelve of them from sub-Saharan Africa. In order to see peoples with lighter skin than expected we must move out of sub-Saharan Africa. Fourteen such populations were found, including some in North African countries, and in Iran, Saudi Arabia, Israel, Nepal, Tibet, Cambodia and China.

Things become more confused when we move to the New World. Follow the Equator and go west from Africa to the north of Brazil, close to the course of the Amazon. In the north-east of Brazil there are many black people, the descendants of West Africans transported in the Atlantic slave trade. However, the indigenous peoples of this area, the Cocama, Tocuna, Cubeo, Yanomamo and others, cannot be described as black. The sun, which has apparently forced the evolution of very high melanism in Africa, has had little effect on skin colour at the same latitudes in the New World. Go north through Central and North America and the strong and consistent reduction in skin pigment found in the African-European transect is missing. Indeed, when we reach the Labrador Eskimo group – the Greenland, Baffinland and Netsilik Eskimos – skin pigment suddenly becomes quite dark. Jablonski and Chaplin's data show that southern Greenlanders have much lower skin reflectance than expected from a consideration of UV intensity in this area of the far north.

A simple correlation between UV intensity and skin colour is not to be found anywhere in the world, and certainly not in the New World. One possible explanation is nutrition. A diet rich in vitamin D, the argument goes, requires no loss of pigment – hence eating lots of fish and marine mammals may mean that Eskimos have enough vitamin D despite their dark skin. It may then make sense to retain a dark skin in order to protect against very occasional high levels of UV. There is also the possibility that many populations in the Tropics, particularly New World populations, have simply not had the time to acquire a dark skin. The ancestors of American Indians are thought to have reached the northern areas of the New World about eleven thousand years ago, and their arrival in Central and South America must have been even more recent.

Are there advantages and disadvantages to skin colour that are powerful enough to explain the current distribution of pigmented skin? Although melanoma is often an aggressive and life-threatening skin cancer, it usually manifests itself in middle age, long after most people have had their children, and so is not likely to be a strong selective force for the evolution of pigmented skin. It is true that black skin tends to reduce vitamin D synthesis and that lack of vitamin D may lead to rickets.[5] However, the selective importance of even this has been disputed. For example, Robbins[6] has pointed out that vitamin D may be stored in muscle and fat, and therefore even a short summer will afford enough for humans to survive throughout the year. In Robbins' view vitamin D deficiency does not occur because of dark skin colour, but as a result of urbanisation, overpopulation, poor diet and insufficient opportunity to be outdoors.

There are counter-examples to the lack-of-time explanation, too. Jared Diamond of the University of California has pointed out that Scandinavians are relative newcomers to the north of Europe.[7] This area was covered by an ice sheet up to about nine thousand years ago, and colonisation occurred about five thousand years ago. Despite this short period of time, most Scandinavians have acquired a light skin. Presumably pigment loss occurred much further south than their present position. More puzzling still are the native Tasmanians, who were dark-skinned despite living

42° south of the Equator. They seem to have had enough time to develop light skin. Their island has been isolated from the mainland for ten thousand years and they lacked ocean-going boats, so a recent migration from the tropical north was not possible. Unfortunately we cannot determine their skin colour by modern techniques because they succumbed to disease and persecution in 1876. However, we do know that their customs included the common practice of the capture of women for marriage, and that they occasionally practised polygyny.[8] This suggests that the Tasmanians had quite strong sexual selection, and that as a consequence pre-natal testosterone was high and disease resistance low. In such a society a barrier to the proliferation of bacteria and fungi in the skin affords a strong selective advantage. There is evidence that melanin provides such a barrier. This presents a different hypothesis, in which dark skin is related to low disease resistance and high pathogen levels. If this is correct, there should be strong evidence for a link between melanin and disease.

A number of authors have suggested that skin colour is a correlate of disease resistance or susceptibility. However, the case for the anti-microbial properties of melanin has been best and most recently put by James Mackintosh.[9] Melanin is found throughout the animal kingdom, and is so widely distributed among so many diverse groups that mechanisms for its synthesis must have arisen as much as 500 million years ago. The system of biological catalysts or enzymes that bring about the production of melanin is known to have powerful anti-microbial effects in a range of animals. In humans and other vertebrates, Mackintosh has suggested, melanosomes and melanin of the skin form a barrier which inhibits the invasion of bacteria and fungi through the outer surface of the skin, and reduces the growth of micro-organisms when the basal layers of the skin are breached. Melanin will therefore be most common in those tissues that characteristically have high numbers of micro-organisms, or are exposed to mechanical abrasion which aids pathogen entry, or tend to lie outside the effective surveillance of the immune system. The skin is one obvious tissue that possesses these characteristics. The outer part of the skin is subject to sunburn and often abraded; wounds are common, and entry of micro-organisms is difficult to oppose because much of the blood supply lies below the base of the epidermis. Melanin occurs throughout the skin but characteristically shows a gradient, so that invading bacteria entering the skin encounter higher melanin content as they penetrate further. The greatest concentration of melanin lies at the base of the outer covering or epidermis of the skin. This then seals the blood and nutrient-rich dermis from exploitation by bacteria.

If we view melanin as an anti-microbial, rather than a protection against UV, some curious facts about its distribution become understandable. For example, nocturnal animals such as the insectivorous and fruit-eating bats and the possum have pigmented skin. Protection against the sun is of little use in nocturnal species, but a defensive shield against bacteria and fungi remains an important and worthwhile adaptation whether one is active during the day or night. Similar reasoning can be applied to the detailed distribution of melanin on the surface of the skin and to its remarkable abundance in internal tissues. In babies the genitalia and the area around the anus may often be more heavily pigmented than the shoulders and arms. Melanin is also found in women's nipples and tends to accumulate in higher concentrations during breastfeeding. This is to be expected if abrasion may lead to infection. Within the body, away from the effects of UV, melanin is also plentiful. The inner ear, the eye, some tissues in the brain and the membrane that lines the abdominal cavity may be richly supplied with melanin in many animal species.

Melanin acts as a mechanical barrier to the entry of micro-organisms, that the action of the melanosomes may be more complex. Central to the function of the immune system is the process of phagocytosis, in which specialised cells engulf bacteria and fungi and then fuse with vesicles called lysosomes. Within the lysosome is a rich soup of enzymes and substances such as nitric oxide which kill the micro-organism. Melanosomes are essentially melanin-rich lysosomes. Once inside the melanosome a bacterium is trapped within the thick melanin polymer and killed by the action of a range of toxins. But what of the melanocyte cells which produce the melanosomes? They are active in engulfing and killing bacteria, but their action may go further, for there is evidence that melanocytes cut out the characteristic 'molecular identity cards' of the micro-organism and present them on their cell surface. Once there, the immune system is able to identify the essential elements of the bacterium and mount an effective response throughout the body.

If these arguments are correct, very black people should be more resistant to skin penetration by bacteria and fungi than lighter-skinned people. There is some evidence to support this. Let us consider the relationship between skin colour and HIV and AIDS. HIV is usually contracted during sexual intercourse. If melanin is a barrier in the skin to bacteria and fungi, why should we expect to find a relationship between skin colour and the HIV virus? HIV is fragile – transmission from an infected individual to a non-infected partner is by no means certain and may require many sexual encounters. In males there is evidence that HIV is often acquired via the foreskin of the penis. Thus circumcision appears to protect against infection, and rates of HIV/AIDS are lowest in those groups in which male circumcision is practised.[10–13] However, when the skin of the penis or vagina is breached by fungi or bacteria the probability of HIV infection is increased. Melanin in the foreskin or the vagina may protect against genital sores and therefore HIV infection. It is known that the foreskins of black men contain substantially more melanin than those of white men.[14] Is such a barrier effective against HIV penetration? With my colleagues Peter Bundred and Peter Henzi I have investigated this possibility by comparing skin colour and HIV/AIDS rates in countries south of the Sahara.[15] We found the lowest rates of HIV/AIDS in those countries in which the indigenous peoples were very black, and the highest rates of HIV/AIDS among lighter-skinned peoples. In other areas of the Old World, where skin colour is much lighter, there was no relationship between melanin and HIV/AIDS rates. The fact that sub-Saharan Africa contains some of the blackest groups of people and has the highest rates of HIV/AIDS in the world suggests that

melanin is not sufficient in itself to stop the spread of HIV. There
appears to be little doubt that the present epidemic is driven by
high rates of unprotected sex between men and women: in such a
situation melanin may give vital extra protection. This finding, if
correct, may be of public health importance. In even the blackest
African groups there is variation in skin colour. AIDS-related
natural selection against lighter-skinned people will result in a
drive towards darker skin colour. However, we could intervene by
targeting health education and free condoms towards relatively
lighter-skinned people within these African populations. This may
be an inexpensive and effective method by which to slow the spread
of the HIV virus in sub-Saharan Africa.

This theory of melanin allows us to view black skin in a new
light. It means that melanin secretion may arise as a response to
immune suppression such as that which, as we have seen, may
be associated with testosterone (p. 72). Now we have the possibility
that sex, and more precisely strong sexual selection acting through
polygyny, may explain much about the distribution of black and
white skin.

An overemphasis on melanin as a response to UV has obscured
the relationship between sex and skin colour. In all known human
societies females are lighter-skinned than males.[6] Both boys and
girls grow darker as they age, but at puberty the skin of girls grows
lighter while that of boys continues to darken. Women also tend
to show cyclical change in skin colour, growing lighter when
ovulating and becoming darker at the beginning and end of their
menstrual cycles. This relationship between skin colour and fertility
can also be seen later in life as women grow darker after their
menopause.[16]

Sex-related differences in skin colour correspond to the relative
amounts of oestrogen and testosterone found in women and men.
Oestrogen reduces or has little effect on melanin production,[17] while
testosterone increases it.[18] This applies not only to melanocytes in
human skin, but also to those found in hamsters, rats and toads.[19]
High oestrogen and low testosterone are correlates of fertility in
women. Thus unsurprisingly there is evidence that, within the

normal variation of the population, men in many cultures have a preference for females with light skin. Van den Berghe and Frost have shown that in a sample of fifty-one societies, forty-seven showed preferences for light skin in women.[20] Women show no such strong preferences for colour in men.

Light skin and attractiveness in women cannot be explained away by class differences arising from colonialism. Doug Jones[16] has pointed out that in ancient Rome women often lightened their skin with cosmetics. This practice continued despite the Roman conquest of many northern tribes where very light skin was commonplace. Thus the attraction of light skin was probably not social position but the promise of fertility. Light skin may also be more attractive because it is associated with an upper-class habit of spending less time outdoors. However, Jones points out that in societies where such class differences do not exist there remains a male preference for light-skinned women. In the eyes of men, light skin means high oestrogen, youth and fertility in women. If melanin is related to sex-hormone levels, can we see a relationship between skin colour and marriage systems which may influence the relative levels of testosterone and oestrogen?

In the previous chapter I argued that polygyny selects strongly for male fertility and competitive physical ability. As a result testosterone levels, particularly prenatal levels, are increased. Is there evidence that polygyny is related to skin colour? In his *Ethnographic Atlas* George Murdock lists 862 human societies.[8] He then groups these into 412 'clusters' made up of very similar societies which share a recent common history. A consideration of the marriage system and latitude of these clusters provides us with clues as to how sex influences skin colour.

First, it is necessary to select one society at random from each cluster. This prevents clusters with large numbers of societies from dominating our results. Second, we must quantify the influence of sex. We are most interested in a correlate of increasing male competition for fertile women, for it is this that will drive an increase in testosterone. A scale of increasing male competition would work as follows:

(i)   polyandry – women are allowed more than one husband
(ii)   monogamy – men have one wife at a time
(iii) limited polygyny – men may have more than one wife at a time
       but this arrangement tends to be relatively rare
(iv)   sororal polygyny – men choose their second wife from
       their sisters-in-law but wives do not live with their husbands
(v)   sororal polygyny with shared quarters – wives live with their
       husbands
(vi)   polygyny – not restricted to wife's sisters; wives occupy separ-
       ate quarters
(vii) polygyny with shared quarters – wives live with their husbands

Where wives occupy separate quarters to their husbands the
probability of cuckoldry – which tends to break down the effects of
polygyny – is increased. In such systems a few men may appear to
have exclusive access to many wives, but clandestine relationships
will tend to reduce the certainty of paternity. Polygynous societies
where wives occupy their husband's quarters (levels v and vii) may
therefore show the signs of intense male competition, i.e. high
testosterone levels, but residence away from husbands is likely to
reduce this effect. Furthermore, non-sororal polygyny often
features capture of females for wives. This is likely to include fights
between males, and places strong selective pressure on such things
as strength, cardiovascular efficiency and spatial judgement.

   We are now in a position to examine the relationship between
strong polygyny (levels vi and vii) and latitude. Black skin is most
common near the Equator, but this is not true of all the continents.
Africa is where we find groups with the blackest skin, and also those
peoples who are blacker than expected from a consideration of
UV alone. In our total sample of 412 societies there are 122 or 30%
that are strongly polygynous. Considering the Tropics only, this
proportion increases to 37% (ninety-eight out of 266 societies).
When we consider African peoples the amount of polygyny
dramatically increases. Within the Tropics we find that 78% of African societies (sixty-six out of eighty-four) are strongly polygynous.
Moving closer still to the Equator (10°N to 10°S) the proportion of

African polygyny increases further, to 85% (fifty-five out of sixty-five) of groups. As we have seen, the latitudinal change from black skin to white is seen best along a line from the tip of Africa to northern Europe. Therefore I will consider the type of marriage systems found along this transect, from 55°S to 70°N. Within 20° north and south of the Equator strong polygyny is present in 76% of our sample, between 21° and 40° the proportion reduces to 27%, and between 41° and 70° it is 8%. Thus proximity to the Equator, black skin and polygyny are closely linked, while distance from the Equator, white skin and monogamy are the norm. However, this applies only to the Africa–Europe transect. In the New World, strong polygyny and black skin are rare, and polygyny is not most common near the Equator. Thus the proportion of strongly polygynous societies with latitudinal change is 16% in the 0° to 20° band, 11% between 21° and 40° and 23% from 41° and beyond. A marked association between latitude and black skin does not exist in the New World, possibly because polygyny is not related to latitude. Indeed, it is quite common in northern Canada and Greenland, which might explain why Eskimos are much darker than we would expect from a consideration of UV intensities alone.

It seems that we now have two major influences on skin colour. Very black skin is related to high UV intensity and strong polygyny, and very white skin to low UV and monogamy or polyandry. But can we separate these influences? The skin-reflectance data of Jablonski and Chaplin afford us this opportunity. Their sample provides skin-reflectance values from indigenous people in eighty-five populations, only four of which are from the New World (two from Brazil and two from Peru), the remainder from Africa, Eurasia, the East Indies and Australia. If we were to assume that skin colour was determined solely by the need to protect against UV light, the average annual UV experienced in an area should allow us to calculate the expected reflectance of the skin for that population – low (i.e. black skin) in the Tropics and high (i.e. white skin) at high latitudes. Although Jablonski and Chaplin found that observed skin colour was close to expectations, an inspection of their data shows that many polygynous African populations south of the

Sahara are blacker than expected and others, such as the poly-androus Tibetans and Nepalese, are lighter than expected. Is the nature of the marriage system driving these exceptions towards dark or light skin?

Figure 7.1 shows that light skin is found in societies in which polyandry and monogamy are the norm. Black skin is found in societies which practise polygyny. It is possible, using the appropriate statistical techniques, to determine from these data whether it is UV light or the marriage system that is strongly associated with skin colour. It turns out to be both. Melanin of the skin, we can conclude, protects against UV light and the invasion of micro-organisms. Furthermore, melanin's function as a barrier against disease is most important in polygynous groups where testosterone lowers immunity levels.[1] Can we take this further and use finger ratios to shed further light on the melanin/UV/testosterone associations?

We know of some differences in adult testosterone levels between groups. For example, studies have shown that Bushmen of Namibia have lower testosterone concentrations than the Kavango of the Namibian and Angolan border area.[21] Bushmen are also far paler than the Kavango, as we would expect if testosterone was an important factor in the evolution of skin colour. Similarly, African-Americans show consistently higher adult levels of testosterone than white Americans. The differences range from 15 to 20% and are unlikely to be caused by class differences or lifestyle factors such as obesity and smoking habits.[22,23] Apparently a West African ancestry is associated with high testosterone concentrations.

There has been one study of the relationship between individual skin colour and finger ratio.[1] The sample was 230 white subjects from the north-west of England, comprising 115 men and 115 women. Skin colour was measured using the 'PocketSpec Bronz' device, which gives a colour score between 0 and 1,000. On this scale good-quality white paper gives a reading of about 30 and black objects a reading of approximately 900. Skin colour was recorded from the inner surface of the upper right and left forearm. The average colour score was 662, and as expected males were darker

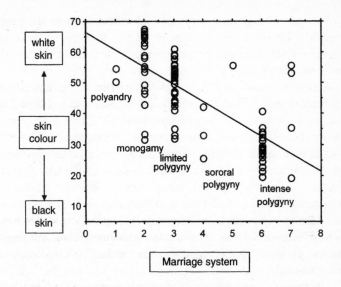

FIGURE 7.1
Marriage systems influence skin colour. This graph illustrates the effect on skin colour of increasing competition among men for wives. Marriage systems such as polyandry and monogamy are associated with weak competition for wives, low levels of early testosterone, high immune competence, female-type finger ratios, and little need for an anti-microbial barrier (melanin) in the skin. These are white-skinned peoples. Very black-skinned peoples have intense polygyny, which often includes capture of wives, and is associated with high levels of early testosterone, low immune competence, 'Casanova-type' finger ratios, and a need for melanin in the skin as an anti-microbial barrier.

(average 667) than females (656). An inspection of the relationship between finger ratio and skin colour showed that, while there were no associations between skin colour and finger ratio in males, in females a 'Casanova-type' finger ratio was associated with a relatively dark skin, while a high finger ratio was found in light-skinned women, indicating that high oestrogen before birth is related to light female skin.

Little is known about population differences in early testosterone concentrations. The finger ratio now affords us an opportunity to compare average ratios across populations at different latitudes. I have considered the limited data (from ten populations) we have thus far.[24] There is indeed a tendency for a low finger ratio to be found among peoples in the Tropics. As we move to higher latitudes the dominant finger pattern becomes that of a female-type ratio. We must be cautious about these data because we need to measure more populations; however, comparisons of finger ratio between ethnic groups support the idea that in the Old World, prenatal testosterone is often high in the peoples near the Equator and tends to reduce as we move to higher latitudes.

High testosterone before birth appears to be inextricably linked with strong competition among men for women and to dark skin. Is it also linked to men's physical prowess and athleticism?

VIII

Fingers, Running Speed and Football Ability

Suppose I were to say to you that measurements of finger length could predict athletic and sporting ability in men, and that, since the finger ratio is fixed early, boys with the potential to acquire running and skiing speed and endurance, and to excel in sports such as football, could be identified and their abilities nurtured and developed. It could be argued that this would diminish the excitement and uncertainty of sport; what is certain, however, is that it would give us the opportunity to understand some of the important factors which underlie athletic and sporting ability, and to identify great natural champions. Let us begin by looking at the facts concerning finger ratios and sporting ability.

Men's running speed, skiing speed, football ability and probably more are strongly predicted by the 'Casanova-type' finger ratio. As we have seen, finger ratios may tell us something about the efficiency of the heart and the blood vessels. By extension, finger ratios should also tell us something about heart function in healthy men.

First, let us consider sex differences in athletic and sports abilities. Take, for example, the world records for track and field events. In track disciplines the male 100 m record in 2002 was 7% faster than the female, 10% for the 200 m, 11% for the 800 m, 10% for the 1,500 m, 13% for the 5,000 m, 12% for the 10,000 m and 9% for the marathon. The average sex difference in these running speeds is about 10% in favour of men. Even larger differences may be seen in some of the field disciplines, with male advantages of 15% and 16% in the high jump and long jump. In some events there has been a narrowing of the sex gap in performances as more women compete, but the difference has stabilised for most disciplines. Of course, in the general population there is overlap between the sexes in these performances. The best women can run faster than many men. However, at each level of ability – world, national and regional competition – the best men can run faster, jump higher and further and are stronger than the best women.

These sex differences probably arise as an evolutionary response to direct competition among men for women or for resources that attract women. In other words, the sex differences in strength and speed have their roots in what is known as sexual selection. If I am right, we should see that the finger ratio predicts abilities which are useful in fights between men – strength for grappling with an opponent, the judgement of distances and shapes for striking opponents, precision in kicking and punching, the precise use of clubs or bats, running speed in pursuing or fleeing from opponents. Excellence in such pursuits should be predicted by 'Casanova-type' finger ratios.

If finger ratios are associated with running speed, endurance and visuo-spatial judgement, it is likely that, as I have suggested, pre-natal testosterone and oestrogen have effects on the muscles, heart, blood vessels and brain of the developing foetus. With this in mind I shall now examine relationships between finger ratios and sporting proxies for male fighting ability.

As men tend to be stronger than women, it follows that low finger ratios should predict strength. There has been one unpublished study thus far of the relationship between finger ratio and strength.[1] The sample was fifty-two university undergraduate males with an average age of twenty-one years. The advantage of such a sample is that the subjects were extremely unlikely to be taking steroids to enhance muscle size and strength. Their strength was tested using three basic lifts – the shoulder press, the bench press and the squat. The shoulder press tests the strength of the deltoids or shoulder muscles which attach to the upper arm and the shoulder girdle, the trapezius which runs from the deltoids to the neck and gives the 'slope' to a weightlifter's shoulder girdle, and the triceps which straightens the arms. Strength in the bench press is dependent on the powerful pectoral muscles of the chest and the frontal part of the deltoid muscles. The squat tests the strength of the massive thigh and buttock muscles.

A weight was chosen which the subjects were confident they could lift for at least five repetitions. The exercise was performed to failure or to ten repetitions, whichever was reached first. Dumbbells were used for the shoulder press and were lifted from the shoulder to the overhead position. The bench press and squat were performed with a barbell. The measure of strength which was adopted was the weight lifted multiplied by the number of times it was lifted (thus a participant who lifted two 20 kg dumbbells for ten repetitions in the overhead press was judged to have lifted a total of 200 kg per arm). The subjects were students who were merely interested in increasing their strength and fitness. The average total lift per student was therefore fairly modest, at 137 kg per arm in the overhead press, 482 kg in the bench press and 785 kg in the squat.

The strongest association was between the finger ratios of the right hand and strength in the overhead press (Figure 8.1). Here the subject with the lowest ratio (0.91) lifted in total 175 kg. In comparison, the participant with the highest ratio (1.01) lifted 65 kg. Of course we must consider other factors when assessing strength. For example, it is common sense that heavy men can on average lift greater weights than light men. However, it is possible statistically to remove the effect of body-weight differences and examine the influence of finger ratio alone on the weight lifted. When this was done it was found finger ratio remained an important predictor of strength in the overhead press. Of the remaining lifts there was a similar association between masculine finger ratios and strength in the bench press, but this time the relationship was weaker and it was with the left hand. There were no associations between finger ratio and strength in the squat, although this does not rule out the possibility of an association between finger ratio and squat poundage in elite-strength athletes rather than amateurs interested in small strength gains.

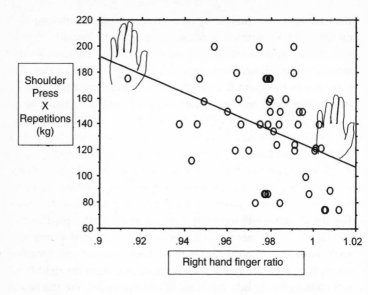

FIGURE 8.1
Men with a 'Casanova-type' finger ratio tend to lift more weight in the shoulder press than men with a female-type finger ratio.

This preliminary study has provided some valuable information, but it should be repeated using drug-free, elite-strength athletes and heavier weights. It may be that finger ratios can be predictors of strength potential in young athletes in such disciplines as the Olympic lifts, power lifting and shot putting. However, my overall impression is that 'Casanova-type' ratios are more strongly associated with running speed than they are with strength.

The times of sprinters, middle-distance and long-distance athletes all appear to be related to their finger ratios. We have less information on sprinters than on middle- and long-distance athletes; in the last category, however, the association between finger ratios and speed appears to be strong.

I will illustrate these associations with reference to three samples of runners.[2,3] The sprinters consisted of forty-seven students aged eighteen. Their times for 100 m were recorded in two races and then averaged. The middle-distance runners were seventy-one club athletes with an average age of twenty-three years. The athletes had completed timed 800 m and 1,500 m races. Of the total sample there were forty-six runners who were able to supply times of their last three 800 m races and thirty-nine for their last three 1,500 m races. An average time was calculated for each athlete. Finally, long-distance runners in two inter-university races were measured. The competitors included fifty-seven athletes in a race over 2 miles near Manchester and forty-eight in a 6-mile run near Edinburgh.

All three groups showed associations between their finger ratios and their race times and/or finishing positions. In general, athletes with low ratios ran faster than runners with female-type ratios. For the sprinters the right-hand ratio was the better predictor of fast times. The runner with the lowest ratio (0.93) had a time of 13 seconds, while the competitor with the highest ratio (1.00) returned a time of 15 seconds. In the middle-distance sample the right-hand ratio was again the best predictor of running speed. For the 800 m runners the athlete with the lowest ratio (0.93) returned a time of 1 minute 55 seconds, while the runner with the highest ratio (1.06) ran 2 minutes 35 seconds. The same athletes gave average times of 3 minutes 47 seconds and 4 minutes 50 seconds respectively for the 1,500 m. In the long-distance races the right and left hands were both strong predictors of the finishing positions of the athletes. The first three finishers in the Manchester race had ratios of 0.93, 0.95 and 0.96, while the last three all had ratios of 1.00. Similarly, the winners of the Edinburgh race (Figure 8.2) had ratios of 0.92, 0.94 and 0.95, while the runners who finished in the last three had ratios of 0.98, 1.00 and 1.01.

Strong as the finger ratio/time associations were in the long-distance runners, there was something else that was stronger. The successful runners trained with mind-boggling ferocity and regularity. It was the number of their training sessions that primarily determined the running speed of these athletes. It could also be seen that the frequency of their training sessions was associated with their finger ratio – that high prenatal testosterone, as indicated by a 'Casanova-type' finger ratio, enables a long-distance athlete to

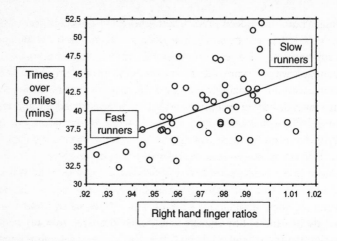

FIGURE 8.2
The running times of forty-eight athletes who competed in a 6-mile cross-country race. The athletes with 'Casanova-type' finger ratios generally returned faster times than those returned by runners with high ratios.

train regularly and with great intensity, thus conditioning his body to speed and endurance across great distances. This would be due to the greater efficiency of the hearts of male athletes with low ratios.

Track running is a fairly straightforward discipline, with simple demands. Slalom skiing is more complex. Speed downhill is essential, and in common with running, time is the measure of performance. However, the competitors have to negotiate a series of gates. At speed this requires precise judgement of distances. The event places demands on the whole body, particularly the large muscles of the thighs and buttocks, the cardiovascular system and those parts of the brain that are concerned with visuo-spatial perception. We should therefore expect that a 'Casanova-type' finger ratio will be predictive of potential in slalom skiing.

I conducted the one study thus far of the relationship between skiing times and finger ratio on a wet July day in Rawtenstall, Lancashire.[4] Each competitor completed two runs, and their faster time was used as a measure of performance. It turned out that finger ratio did indeed predict skiing times. The average age of the fifty-two male and twenty female skiers measured on the slope that day was twenty-one years, but this hides an immense variation – the youngest was six years old and the oldest, the president of the local club, was an amazingly youthful seventy-two years. However, age takes its toll on everyone, and I therefore restricted my analysis to the fifty-seven skiers in the six- to twenty-five-year age group. Within this group there was the usual sex difference in finger ratio, with males having lower ratios. However, the average ratio for the skiers as a whole was particularly low (0.96 for the right hand). Comparing this with the right-hand ratio of males and females from the local population, I found that it was indeed lower than would be expected by chance alone (0.96 compared to 0.99), indicating that people who take up skiing have more masculine finger ratios than the average values found in the population.

Skiing requires practice, and so before looking at the times we must remove the influence of experience, age and sex. The remaining variation in times was associated with finger ratio. For example, the ten fastest skiers had an average right-hand ratio of 0.94, lower than the overall average for the sample of 0.96. The average for the ten slowest skiers was higher than the average, at 0.98. This was only a first attempt to look at finger ratio and skiing speed. It remains to be seen whether the variation of times in elite skiers of similar age and experience can be predicted by finger ratio, and whether any effect will be more powerful in speed skiing, slalom or endurance races.

I now turn to football, and here I must declare a personal interest. I grew up in the north-east of England, where most boys dream of captaining the England football team. I was no exception, playing the game every day on local fields and street corners until the light failed and I had to trudge home. I could have made it to greatness and played for my beloved Sunderland (or so I tell my daughters), if it hadn't been for the lure of evolutionary biology . . .

What qualities make a great football player? Obviously the ability to kick and (in the case of goalkeepers) punch a ball accurately and with great force. Add to this speed over short distances, endurance, accuracy in passing, ball control, judgement of distance, courage, strength, commitment to the team and an instinctive ability to rally and inspire a small group of men to coordinated and determined physical action. In short, football can be seen as a metaphor for physical combat between individuals and small groups of men.

There are two published studies on finger ratios and football.[3,5] Both address the fundamental problem of measuring ability in what is a very complex sport. A game of football is won by the efforts of a team of players, and so team results are not a good indicator of individual ability. How, then, can we measure individual performance? In such a situation it is best to take a number of approaches. We can simply ask players to rank their overall ability against some sort of scale which starts at, say, 'enthusiastic but not gifted' and extends to a level of ability which would be 'appropriate for membership of national teams'. In order to tackle the problem of subjective responses we can then ask others, such as their fellow players, their captain or their coaches, to judge their level of performance. Alternatively, we can attempt to break down ability into such things as speed, endurance, passing and ball control. Traits such as speed may be measured quantitatively, while others are again a matter of judgement.

Let us now consider these and similar approaches taken in three studies I conducted with Rogan Taylor.[5] First, 128 young men were asked to rate their ability at their favourite sport on a scale from 1 to 10. The rating scale ran from social involvement only (1), through participation in organised sport, to county, national and international level (10). All participants were amateur sportsmen, including runners (45%), football players (14%) and martial arts competitors (10%). In the results, men with 'Casanova-type' finger ratios of 0.92 to 0.94 reported high sporting achievement. Those with female-type ratios tended to rate themselves as poor sporting achievers. Note that this sample shows a bias towards sports that place cardiovascular demands on the participant (running and football).

Second, the relationship between finger ratio and visuo-spatial judgement was assessed. Football involves fine judgement of the surface of the ball and the part of the foot which makes contact with the ball. Strike a football obliquely and it will fly away at an angle from the kicker; vary the orientation of the foot to the ball and you can cause the trajectory of the ball to change in flight. All this requires great skill if it is to be done in a controlled way, and may be aided by accurate visual perception of shapes. A sample of 125 young men were tested for their visuo-spatial abilities using 'mental rotation' diagrams. The task required pairing up identical shapes that were shown in a rotated position. It was found that participants with 'Casanova-type' ratios scored higher in these mental rotation tests than men with female-type ratios.

Third and last, finger ratios were measured in a sample of 304 professional football players and 533 young men who did not play football professionally. All the footballers had played or were currently playing in English professional football in the 1998–9 season (of course, this is not to say that they were all English). Within the sample of football players there were players from two clubs in the Premier Division, Liverpool and Coventry City. Division I was represented by Sunderland and Tranmere Rovers. Division II players were provided by Preston North End and Oldham Athletic, and Division III was represented by players from Cambridge United and Rochdale. Former distinguished players often serve as coaches at professional clubs. We were fortunate enough to measure twenty-one of these, including great defenders such as former Liverpool players Phil Thompson and Ron Yeates and Derby County's Roy McFarland, and forwards Peter Reid and Kevin Sheedy of Everton. Finally, during a memorable evening, we measured the fingers of twenty-nine of football's greatest players.

The occasion was the centenary celebrations marking the one hundredth English League Championship. The guest list reflected the depth of talent in the English Football League, and included one hundred 'League Legends' to mark a century of English football. The earliest of these greats were being honoured posthumously, but there were still many great players I had watched over the years, and many whose reputation had survived in the minds of those who had endlessly discussed the relative merits of their skills. All were great football players, but the qualities they showed on the field were many and various. What about their finger ratios?

There were great differences between the average finger ratio of our large sample of non-players and the groups of footballers. The average for the former was 0.98, while for the latter it varied between 0.94 and 0.955. Among the young professionals the average ratio varied little between the divisions, although one problem we experienced in assessing the data is that clubs move up and down. Thus of those in the original study, Coventry have since moved down to Division I (now called the Championship), Sunderland have moved up to the Premiership and returned to Division I, Tranmere have moved down to Division II (today's League One), Preston have moved up to Division I and Cambridge out of league football altogether. However, considering the clubs as a whole, we found that on the whole first-team senior players had more extreme 'Casanova-type' ratios than reserves. One possible explanation of this finding is that competition for first-team places identifies the gifted masculinised players, and this shows through in their lower finger ratios. Youth-team players' finger ratios were generally intermediate, higher on average than senior first-teamers and lower on average than reserves; this is as I would have expected given that they comprise a mixture of potential first-team players, reserves, and those who drop out of the game altogether. Also as expected, the coaches and 'League Legends' were the most masculinised players in the entire sample, with average ratios close to 0.94. Selection for the national team imposes a further 'pressure' to identify masculinised players. In the total sample there were thirty-seven players who had played for one of the 'Home' countries (England, Scotland, Wales, Northern Ireland and Eire). The internationals had a lower finger ratio (0.94) than the average for their colleagues without international experience (0.95), and there was even a tendency for players with many international appearances to be particularly masculinised in their finger ratio.

All these observations and conclusions depend on the judgement of coaches, international selectors and others. In a study of seventy-one amateur English players I conducted with Peter Bundred and

Rogan Taylor,[3] an alternative approach was adopted. The ratings of each player's ability were provided by the players themselves and by five of their team-mates. There was strong agreement between self-perceived ability and the judgement of others. More importantly, both forms of rating were associated with the finger ratios of the players. Men with 'Casanova-type' finger ratios had high football ability, whether this ability was judged by the player or by his team-mates.

Our first study of football players in English leagues provided evidence that finger ratios were predictive of football ability. Surprisingly, and for reasons I do not yet fully understand, most of these relationships were strongest for the finger ratio of the left hand. We observed the same tendency in our second study, performed in the most successful of all football nations, Brazil.

Brazil is often referred to as the true home of football. The country has never failed to qualify for the World Cup Finals, and has won the trophy a record five times, the last triumph in Japan in 2002. How do we account for Brazil's success? A glance at the organisation of the game there reveals a system that tolerates and even encourages too many games at all levels, exacting a high toll of injury and stress on players. It is certainly not the administration of the game, then, that lies at the heart of Brazil's dominance.

In collaboration with Rogan Taylor and senior coach João Paulo Medina, I was recently fortunate enough to measure the finger ratios of a Brazilian First Division side, Sport Club Internacional ('Inter' for short) of Porto Alegre.[3] The hands of ninety-nine players were photocopied, of whom thirty-three were members of the senior playing staff. This group of players had an average left-hand finger ratio of 0.93, more masculinised than English football's 'League Legends'. The first-team squad consisted of twenty players with an even lower average finger ratio of 0.92, and thirteen reserves with a mean ratio of 0.96. Considering the first-team squad, six players had astonishingly low ratios of below 0.90 (see Figure 8.3, which illustrates the left hand of one such player with a ratio of 0.88). This group included a forward with a ratio of 0.885, who was subsequently transferred to Barcelona for a fee reported to be in

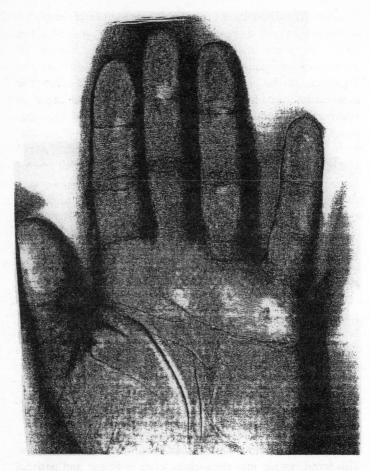

FIGURE 8.3
The left hand of a Brazilian football player. Note the extremely long ring finger relative to index finger in this talented midfielder. The finger ratio in this instance is a very masculinised 0.88 (second finger 75.20 mm divided by fourth finger 85.15 mm).

excess of $12 million. Inter's coaches were asked to rate the players on scales from 1 to 4 (1 indicating highest ability) on endurance, pace, passing and dribbling. A composite score of abilities per player was obtained by summing the ratings for all four qualities. We found that the players with low finger ratios, particularly left-hand ratios, scored highest on this measure of overall ability. Considering each of the four traits individually revealed that finger ratios were most closely related to ratings of endurance. This study again shows that finger ratio appears to be predictive of football ability. Whether endurance is the component of football ability that finger ratio predicts best remains to be seen. However, I have a strong feeling that the ratio probably correlates with not one but rather a multi-dimensional aspect of football ability.

We have discussed ethnic differences in foetal testosterone, and have observed that differences in finger ratio exist between human populations. These may to some extent account for the success enjoyed by nations such as Brazil.

The English League study, meanwhile, included thirteen black players in our sample, whose mean finger ratio was more masculinised (0.93) than that of their Caucasian colleagues (0.95). Welcome strides have been made in the representation of black players in English league football. Stefan Szymanski of Imperial College London has considered the representation of black players in thirty-nine English professional clubs during the period 1974 to 1993. In 1974 there were only four black players, appearing a total of seventy-seven times for their clubs. By 1993 the number of black players had risen to ninety-eight, and their appearances to 2,033.[6] This trend reaches into the highest levels of league and national football. All recent Premiership-winning sides have included a significant contingent of black players (on one occasion Arsenal fielded nine black players in a match), while the England team has seen a steady rise in black representation since 1978, when Viv Anderson became the first black player to represent the national side.

This trend has little to do with an even-handed approach to opportunities for ethnic minorities, or with a policy of paying lower

wages to black players. Job security is very fragile for managers of
clubs, and so all players are included on merit. This does not mean
that it is sufficient simply to pick a team of players with low finger
ratios and expect them to excel. We must first choose from men
who play the game and are enthused by its challenges. Within this
group 'Casanova-type' ratios in the left hand may point us to the
best players. All other things being equal, the standard of national
football teams is likely to be similarly influenced by the prevailing
levels of foetal masculinisation. I do not have any direct evidence,
but I suspect that the multi-ethnic population of Brazil is highly
masculinised before birth. This, coupled with an intense interest in
the game, may lie at the heart of the country's success. It would
mean that finding very low-ratio boys who are interested in playing
football would be easy.

What does this say for the English population, with its feminine-
type finger ratio, and England's chances of winning another
World Cup? I am still hopeful. The percentage of England's male
population with 'Casanova-type' ratios is low, but motivation and
population base are high. Hence potentially very gifted football
players may be rare, but in terms of actual numbers of children they
can still be found. England may have sufficient young players in
each generation to produce a World Cup team of great talent, particularly if such boys can be identified early and their development
encouraged.

Football is a highly competitive sport, and professional clubs
are trying to identify talented boys earlier and earlier. In this
they encounter a problem which arises from different rates of
maturation. Consider two boys, A and B, aged twelve. Let us say that
A is rated more highly than B by professional coaches. In eight
years' time, when these boys are men, will A still be better than B?
We cannot answer this question with any certainty because we
know their chronological age but not their maturational age. For
example, A may have been six or even twelve months in advance in
his development over B. This would give him an advantage in football ability which was determined by maturation and not by innate
skill. The advantage would vanish and maybe even reverse when

both reached maturity. Thus successful members of England youth teams do not automatically or even often progress to become successful members of the senior team. This is not to say that coaches' judgements are unreliable when the players are twelve and reliable when they are twenty. It is merely that judgements of players' skill at twelve years of age are not good predictors of judgements made when they are twenty. There are two studies that address the use of finger ratio and football ability in boys. They were conducted in the youth teams and football academies of Charlton Athletic of England and Internacional of Brazil.

Charlton Athletic are an English Premiership club with a well-structured policy of developing and encouraging youth players, led by Academy Director Michael Brown. In this study the measurements of finger ratios were made from photocopies of the right and left hands, and the identity of the players was not revealed to the measurer. There were six youth squads, the under 12s squad (sixteen players), under 13s (twelve players), under 14s (sixteen players), under 15s (sixteen players), under 16s (eight players) and under 17s (thirty players). Players' performance was rated in the following way. The under 17s included six players who had represented their country at youth level. It was decided to assume that these six represented the best players. Academy coaches then rated the abilities of the remaining players, and the 'top six' were identified in each squad. Finger ratios were measured from the photocopies without knowledge of the coaches' performance ratings.

If we are to test the finger ratio as a predictor of football ability in boys it is important to be clear about our expectations. We should find that finger ratios will predict coaches' ratings in late-teen boys but not early-teen boys, as only in the former are they reliable predictors of eventual adult ability. With the Charlton Athletic data we asked how many of the top six finger-ratio rated players were in the top six coach-rated group for each squad. For the under 12s and under 13s only one boy in each squad was rated in the top six by finger ratios and also by coaches' ratings. This figure rose to two boys in the under 14s and under 15s, and to four boys in the under 16s and under 17s. In the under 17s group the probability that the finger ratios identified four out of six of the top coach-rated players purely by chance was only one in five hundred. This possibility is sufficiently improbable to reject; it is much more likely that finger ratios are predictive of football ability, and in particular of coaches' ratings of football ability. Thus judgements of football ability by finger measurements and coaches' ratings show increasing agreement as one goes from boys of fourteen to seventeen years. This supports the theory that coaches' judgements of adult ability are reliable at seventeen but not twelve years, while finger ratios reliably indicate adult ability when made at any age.

With the Charlton Athletic youth players the left-hand finger ratios again appeared to be better predictors of football ability than the right. This applied to relationships between ratios and coaches' ratings, and also to 20 m running times. The latter finding was surprising, since running speed appears to be better predicted by right-hand than left-hand finger ratios. This suggests a reversal of the usual side-of-body effects associated with prenatal testosterone – perhaps a peculiarity of gifted football players. As a final observation, the finger ratios of these boys give some hope to those of us who wish England well in international competitions. The average finger ratios per squad varied from 0.92 to 0.94. This indicated very strong selection of low-ratio boys from the general population of high-ratio males.

In the study of youth players from Sporting Club Internacional of Brazil, finger ratios were calculated from measurements of photocopies of right and left hands. The sample was twenty-one players of fifteen years and under with an average age of 14.6 years, and fifty-one players of sixteen years and over with an average age of 19.5 years. Coaches' ratings were obtained for pace, endurance, dribbling and passing. Ratings were on a scale from 1 to 4, with 1 indicating highest ability. The scores were then added together to give a form of composite 'football ability' score. The ratings of 6 to 8 points included the seven top-rated players in each group. In the sixteen years and over group there was evidence that on average top-rated players had lower ratios than players with lower coach's ratings, i.e. 0.91 as against 0.94, suggesting that low finger ratios predicted coaches' top ratings for this age group. The opposite picture emerged with the younger group of players. Players with ratings of 6 to 8 points had average finger ratios of 0.94. Those with 9 or more points had a low average ratio of 0.91. As with the Charlton Athletic data, we found that finger ratios and coaches' judgements coincided in players in their late teens and older, but not in younger players.

As before, I interpret these findings as indicating finger ratios as predictors of football ability in adults. They do this at any age, because finger ratios are largely determined before birth. The ratings of coaches are accurate when they are made – they reflect ability at any age – but because ability in young boys is influenced strongly by rates of maturation, these ratings reflect current ability but are not predictive of future adult ability.

In this chapter I have considered male athleticism and its relationship to early testosterone and finger ratio. I have suggested links with physical competition for women because I feel that it is within this framework we can best understand male sport. I now wish to consider male displays to women that do not depend on athleticism.

IX
Fingers and Sexual Attraction

I have shown how our finger ratios are related to our sex, class, ethnicity, family size, personality, predisposition to developmental disorders, serious and infectious diseases, athleticism and sports ability. Many of these things are associated with what might be called our desirability as a mate. A desirable personality might be affectionate, sympathetic, sensitive to the needs of a partner, compassionate and gentle. Women might add certain male-like traits such as self-reliance, assertiveness and competitiveness in the accumulation of status and resources. Men might stress physical attractiveness and youth.[1]

The literature on attractiveness is large and varied. Here I am solely interested in biological traits which signal fertility and good health, and are powerful correlates of attractiveness. Fertility in a sexual partner relates to that evolutionary imperative, the need to reproduce. Good health is a predictor of our ability to provide long-term investment in our children, thus ensuring that copies of our genes are afforded every chance of replicating themselves in order that we may have grandchildren.

Our finger ratio contains clues to the development of the reproductive system – the successful formation of the penis and its associated ducts, and the fallopian tubes and uterus. It also gives clues to susceptibility to major diseases that may cut short our long-term investment in children. It is therefore possible and probably desirable that we pay attention to hands and particularly fingers in our assessments of potential mates, for in the hands lie signals of fertility and good health. However, hands are not always readily accessible, so we may also respond to cues that may be assessed at a distance. These cues may themselves be associated with finger ratios. Let us consider the evidence for these speculations.

Of all the digits of the hand the fourth, the ring finger, is in my opinion the most fascinating. It is the least mobile of all the fingers. If you clench your fist, it should be easy to raise your index, middle and little fingers without moving the others. However, it is difficult and even rather painful to extend the fourth finger from this position. At one level the reason for this is easy to see – the ring finger is supplied with fewer muscles than the other digits.[2] If we consider that body structure and function often go hand in hand, the limited range of movement of the ring finger suggests that it is not mechanically important in manipulation or even in grasping. What, then, is its main function? It may sound bizarre and even laughable, but I believe the ring finger may be a display structure, rather like a modest peacock's tail. Let us begin to consider this suggestion by exploring a few facts.

As I have mentioned (p. 12), this finger is often referred to as the annulus, and we choose to advertise it by the wearing of rings. E. Cobham Brewer in his *Dictionary of Phrase and Fable* (1898) tells us that the Egyptians believed a delicate nerve ran from the fourth finger of the left hand to the heart. In many cultures the seat of love is thought to be the heart, so what better finger on which to wear a wedding ring? Henry Swinburne, in his *Treatise of Spousals* (1680), claims that learned anatomists had found a vein rather than a nerve, the *vena amoris*, which passes from the ring finger to the heart. So is it nerves or blood we are advertising with our wedding rings? It may be neither. Early artists concerned with portraying the human form at its most beautiful and 'advanced' often regarded a long ring finger as primitive and a reversion to an earlier, pre-human type.[3] This conclusion is echoed in comparisons made by anatomists such as Adolph Schultz[4] between human fourth fingers and those of monkeys and apes, comparisons which may be responding to the generally greater volume of hair on the middle part of the ring finger. This relative hairiness is possibly the result of testosterone and its receptor molecule.[5] Receptors in the bone and surrounding connective tissue of the ring finger may respond to testosterone by stimulating growth and the development of hair follicles. If the number of testosterone receptors reduces as one goes from the middle finger to the little finger and then the index finger, this may explain why the ring finger shows evidence of early growth in males which is not seen in females, and why hair growth is most luxuriant on the ring finger and least developed on the index finger.

If my speculations are correct then the ring finger may indeed be a human display trait. Although far more modest in size than the train of the peacock or the feathers of a male lyrebird, and not the focus of elaborate displays to potential mates, long ring fingers are commonly perceived by men and women as attractive and sexy. I shall now discuss the evidence.

What information does a hand possess which might relate to perceptions of sexiness or attractiveness? The length of the fingers is almost certainly related to height, and it is well established both that tall men are more likely to marry than short men, and that tallness is related to large family size. Long fingers, then, may be considered highly attractive by women simply because they are related to tallness in men. Finger length may also be related to the timing of puberty, early puberty bringing about early cessation of finger growth and late puberty favouring long fingers. There is much debate about the factors affecting the timing of puberty, but evidence exists for an association between early puberty and unstable family structures and paternal absence.[6,7] As these factors may reduce a partner's suitability, it could be advantageous to have a preference for long fingers. The weight of an individual may be associated with plumpness of the fingers. People who are obese often have lowered fertility: lean fingers may therefore be seen as attractive by both men and women. We must also consider age. The skin changes its texture and darkens with age. As women's fertility is very strongly dependent on age, it would not be surprising if men perceived the hands of older women as being unattractive. It is highly unlikely that we can look at a hand and calculate its finger ratio in some mysterious, innate fashion, so finger length is probably the key factor in assessing attractiveness.

The question of finger length and attractiveness has recently been considered in an unpublished study I conducted with David Crone. The sample comprised sixty male and ninety-one female undergraduates. Photocopies were made of the palm and the back of the right hand. Ten men and ten women were asked to rate the hands, each photocopy allocated randomly to five of the men and five of the women. The photocopies were rated on a seven-point scale of attractiveness, sexiness, assertiveness and intelligence. Using appropriate statistical techniques it is possible to consider the relationship between, say, ring-finger length and sexiness, independent of the influences of other variables such as height, weight, age and index-finger length. We found that the hands of tall, non-obese men with long ring fingers were rated as most attractive and sexy. Young, slim women with long ring fingers were judged to have attractive and sexy hands. I think it worth emphasising that these features of attractiveness and sexiness were independent of one another. For example, long ring fingers were judged to be attractive independent of the age, height and weight of the owner of the hand. This suggests that the ring finger might indeed be a display trait, for both men and women. It is also worth noting that both men and women recognised long ring fingers as being attractive in both male and female hands.

The relationship between ring finger length, attractiveness and sexiness of the hand has been observed in a similar unpublished study I conducted with Caroline Bradley. However, this time the subjects were men and their hands were rated by women. Furthermore, we differentiated between women who were taking the birth-control pill (about half of our sample) and those who were not. For the latter group we again found a preference for long ring fingers. Those taking the birth-control pill, however, reported a preference for long index fingers, suggesting that the pill might in some way influence women's assessment of male attractiveness in the fingers.

Little is known about how we assess hands during courtship and how important they are in relation to other features such as the face. When we are interacting with a potential sexual partner how much time do we spend looking at their face, body and hands? Do we hold hands to get a 'feel' for finger length in relation to body size, and skin texture and elasticity in relation to age? These kinds of evaluation may be important in our perceptions of attractiveness, but they are only possible when in close proximity to potential partners. In order to hold hands one must first go some way towards admitting an attraction. There may be other ways of assessing hands in potential sexual partners. In this we need advertisements of fertility and good health which are accurately perceived at a distance and are associated with fingers. Music may be such an advertisement.

Music in all its forms gives an enormous amount of pleasure, but why is it universal in the human species? Can we interpret music in an evolutionary framework? At first sight there is no obvious connection between music and natural selection. Charles Darwin stated, 'As neither the enjoyment nor the capacity of producing musical notes are faculties of the least use to man in reference to his daily habits of life, they must be ranked among the most mysterious with which he is endowed.'[8] Music production is costly in time and energy. It does not help us to avoid or kill predators, nor is there any evidence that it increases resistance to infection or directly aids us in searching for food. Music does not help us survive and to adapt to our environment; it appears to be unrelated to natural selection except that it undoubtedly carries a cost.

However, there is another kind of selection, sexual selection, a concept we also owe to Darwin.[8] Humans are not the only species who spend energy and time in wasteful indulgences such as music, nor are such costly ornaments restricted to visual displays. Sound is also important. Darwin realised this and pointed to similarities between human music and the songs of birds, apes and whales. In some animals elaborate ornaments and sound go together in one complex and stunning display – for example, the wonderful train and the characteristic shivering noise made by the peacock. The train grows from the back of the bird, and consists of between 120 and 160 feathers. Many of these terminate in the well-known eye or ocellus structure, but about fifteen to twenty of the longest feathers end in a V shape or fishtail structure.[9-11] The display starts when a peahen appears. The peacock turns his back on her and erects his train. If the peahen does not wander off he then turns, displays the full grandeur of his train and proceeds to 'shiver' the feathers. This expends great energy, and males end their display panting heavily. All the female has to do is remain in close proximity to the male, which he takes as a signal that she is persuaded by his efforts. He then hoots, dashes for the peahen, and finally mates.

If a peacock is not in good condition, he will be unable to bear the cost of the ornament and the display of shivering. The train and the display, then, are honest signals of male condition. Why, though, have males evolved this expensive display when it appears not to increase their 'fitness' for their environment? Darwin recognised the obvious point that trains and shivering have evolved because peahens prefer elaborately ornamented peacocks that can bear the energy cost of train shivering. What this reveals to the peahen is not yet clear, but it may indicate that a peacock is fertile, successful in resisting parasites, has good genes, or some combination of these.

Now consider the possibility that human music is an honest signal which indicates some male characteristic which is of interest to women. Darwin was the first to suggest this, and in recent times it has been championed by Geoffrey Miller of the University of New Mexico.[12] In humans, music is unlikely to be a signal of male condition – writing and making music does not require huge amounts of energy. However, suppose women are interested in cues of male fertility. Their problem is to find a reliable indicator, one which males are unlikely to be able to fake. A possibility is an easily assessed cue for the successful formation of the reproductive system in the developing male foetus. One of the first products of the foetal testes is testosterone that, I have suggested, stimulates the growth of the ring finger. However, this cannot be judged from a distance. The late Norman Geschwind may have found the answer. He suggested that foetal testosterone alters the organisation of the developing male brain.[13–15] More specifically, it promotes the formation of parts of the right side of the brain and inhibits the development of related areas on the left. Musical ability, Geschwind argued, resides in the right hemisphere of the brain. Efficiently functioning testes in the foetus, then, suggest fertility in the adult and right-brain development leading to musical ability. In short, men who make lots of good music make lots of good sperm.

There is some evidence to support this view. A sample of more than seven thousand jazz, rock and classical albums showed that male musicians were ten times more numerous than females, and among these males their output peaked at about age thirty.[12] Although this can to an extent be explained by social factors, such a pattern would be expected with a signal that is advertising the fertility of young and, in the case of classical orchestras, middle-aged men.

Sluming and I[16] conducted a more direct test of the link between music and male sexual signalling on a sample of male symphony musicians. The sample comprised fifty-four men and sixteen women musicians from a well-known British symphony orchestra (the orchestra requested anonymity). As expected, there was a pre-ponderance of men in the sample. The average finger ratio of the musicians turned out to be low, at 0.96 for the left hand and 0.92 for the right. These ratios were considerably more mascu-linised than those of a control sample of local Liverpool men who had an average ratio of 0.98 for both left and right hands. The more masculinised ratios of the musicians could not be explained away by their ethnicity. Most of the orchestra were of white British origin, with the exception of two Scandinavian musicians. After they were removed from the sample the average finger ratios of the musicians remained lower than those of local men. While this finding suggests that male musicians have experienced high testosterone during their development, enhancing their ability to make music, it does not follow that good musicians are also fast runners and good football players. Rather, as a group, musicians have been highly testosteronised before birth, and the effect of the hormone has been particularly pronounced on that part of the brain that is important in the making of music.

With only one sample available to us we must be cautious in our generalisations. There are other possible interpretations of our data. For example, if a long ring finger is in some way directly advantageous in playing instruments, this might lead to accomplished musicians showing masculine finger ratios. I think this is unlikely. The simple mechanical advantage of a long ring finger for playing certain types of instruments is debatable, and we found no substantial differences in the finger ratios of, say, string and non-string players.

The possibility of a direct relationship between finger ratio and musical ability appeared to be strengthened when we discovered that the organisation of the orchestra was reflected in the finger ratios of the musicians. Within each section or instrument group there are positions which are allocated on judgement of musical ability. Some sections had only one musician (e.g. tympani), and for others we had only one measured participant. In sections with two or more musicians there were thirty-nine musicians whose finger measurements had been taken and who were then ranked by the orchestra. We divided these musicians into a 'high'-ranked group (from top-ranked musician down and including the middle-ranked) and a 'low'-ranked group (below the middle-ranked musician). We found that the high-ranked group had more masculinised ratios (average 0.95 in the left hand) than the low-ranked group (average 0.97). Of all the instrument groups the largest section was that of the violins. We had measurements for fourteen ranked male violinists, and found that the best violinists had very low finger ratios. It would appear from this study that there is an association between musical ability and exposure to high testosterone before birth, and that the higher the prenatal testosterone the greater the musical ability.

Returning to the comparison between humans and other animals, there is a striking similarity between animal leks and human bands or orchestras. Communal displays are found in a number of species.[17] The animals congregate on leks or display arenas, where each male has a small territory or court. Bright plumage, loud calls and exaggerated postures are all found on the leks, which are conspicuous and apparently traditional sites to which the animals return over long periods of time. For example, in grouse the leks are found on flat open ground; others species display on the forest floor in clearings with the leaves carefully cleared, while birds of paradise display in the tops of trees. Females visit leks and pick their way around and through the arenas, eventually choosing males for copulation. In such species males contribute sperm but not parental care. It is therefore likely that females seek male features that indicate fertility and/or good genes.

If we now consider performances of human music, some tentative parallels can be drawn. Many popular musicians play or display as a group, often at well-known local venues and in a sexually charged atmosphere. Such comparisons are less obvious in the case of classical concerts. The musicians and audiences tend to be older than at pop concerts. Everyone appears to be serious and engrossed in the complexities of the music. However, all may not be quite as it seems. Classical concerts show evidence of a female-biased sex ratio in the seats close to the stage. Sluming and I[16] have noted that in eleven concerts there were a total of 820 individuals in the first four rows of the centre stalls. Of these 69% were women. In the back four rows of the stalls there were 930 individuals. Of these 51% were women. The excess of women near to the orchestra could not be explained by different admission prices for front and back stalls, since some of these concerts had equal prices across all the seats. Perhaps, then, sexual selection is alive and well in our great concert halls. It may seem far-fetched, and more studies are needed (for example, to determine the proportion of pre-menopausal women in the front stalls), but the genteel world of classical music might benefit from a reassessment.

The data I have discussed in this chapter indicate a close association between fingers and sexual attraction. However, the choice of a sexual partner is a complex thing, too complex to entrust to our conscious mind. We do not understand the function of many of the things we use to assess attractiveness. Our findings regarding the relationships between fingers and our perceptions of attractiveness reveal the richness and complexity that underlies our mate choices. However, thus far I have confined my attention to heterosexual relationships. Can we extend the use of the finger ratio to homosexuality?

X

Fingers and Homosexuality

Reproduction lies at the heart of natural selection, for this is the way of ensuring that copies of one's own genes enter the next generation. Counting numbers of offspring that survive to maturity remains a good, if rather rough and ready, way of estimating fitness. It is important to get the preliminaries to reproduction right, i.e. courtship, partner choice and copulation, in order that one's direct fitness is assured. Homosexuality is therefore difficult to explain in terms of natural selection. It is quite common in both men and women – in Westernised societies about 1% to 4% of men are homosexual, while the rate for women is probably about half that percentage. It is very unlikely that homosexuality arises because of random damage or mutation to genes controlling sexual behaviour, for mutation in general is very rare. It appears to be the case that homosexuals, particularly male homosexuals, have fewer children than heterosexuals. For example, a study by Bell, Weinberg and Hammersmith[1] found that male gays reported about one-fifth the number of children as male heterosexuals. With a very low rate of mutation of genes for heterosexuality to genes for homosexuality, and a high loss of genes for homosexuality through reduced family size, it is difficult to see how homosexuality can be maintained at its present frequency in human populations. One explanation is that humans have now escaped the influence of natural selection, and one result of this has been the spread of homosexuality – however, this argument is undermined by the fact that homosexuality appears to be quite common in other animals.

Rams are famed for their tireless ability to copulate, and the testes of rams are huge in relation to their body size. Under normal conditions rams will cover about five ewes per day. Very large testes indicate the production of vast numbers of sperm. Rams ejaculate about 300 million sperm per mating, and estimates of daily production commonly suggest an average of about 5 billion sperm per ram.[2] These great numbers are needed by rams, though, for they commonly encounter something called sperm competition. This means that when there are two or more rams in the vicinity of a ewe she is likely to have sperm in her genital tract from two or more rams. There is in effect a sperm race going on within ewes. The winners of such races are those rams who produce very large numbers of sperm and who copulate frequently. Successful rams are therefore nature's sperm producers.

However, a ram needs to be more than a super sperm producer; he needs to know with whom to copulate. Mating with ewes increases direct fitness, but copulation with rams wastes sperm. Despite such very strong selection for this most basic form of discrimination, Charles Rosselli at the Oregon Health and Science University in Portland has found that about 8% of rams copulate with other males. More disastrously for their direct fitness, they will only mate with other rams, and so they are exclusively homosexual.[3] It seems that ram and man share similar brain structures associated with homosexuality. In 1991 Simon LeVay found that an area of the brain called the 'third interstitial nucleus of the hypothalamus' or INAH3 differed between gay and straight men. The INAH3 is larger in heterosexual men than in heterosexual women. In gay men the INAH3 is similar in size to that found in heterosexual women.[4] Charles Rosselli has found similar differences in the size of the INAH3 in homosexual and heterosexual rams, and other studies have found similar sex differences in the size of the NIAH3 in male and female mice. Such observations suggest striking parallels between homosexuality in humans and other animals.

Could intensive breeding have introduced mutations into sheep so that this puzzling sexual behaviour is more common than in nature? The answer to this is almost certainly no – such mutations would be quickly eliminated because they would not be passed on.

Moreover, it turns out that homosexuality is found not only in humans and domesticated animals but in many other species of mammal and bird. Bruce Bagemihl[5] has documented 190 species in which homosexuality has been observed, from courting male gorillas and lesbian grizzly bears to same-sex pairs of flamingos. The problem lies in the explanation of homosexuality.

Matters would be simpler, at least for explaining human behaviour, if sexual orientation was simply a lifestyle choice. However, there is accumulating evidence that the factors determining homosexuality, particularly male homosexuality, lie more in our biology than in our lifestyle choices. Male homosexuality runs in families. Brothers of gay men are about four times more likely to be homosexual than the brothers of heterosexual men. This might be connected to the lifestyle adopted within families. However, there is evidence that genetic factors do influence sexual orientation in men.[6]

In this context twins are important, since they experience similar family lifestyles; however, pairs of identical and non-identical twins have different degrees of genetic similarity. Identical twins arise from the same fertilised egg, so they are genetically identical. Non-identical twins arise from two separate fertilised eggs, and therefore share copies of 50% of their genes. If a gay man has an identical twin, his twin has approximately a 40% chance of being homosexual. However, a non-identical twin of a gay man has approximately a 20% chance of being homosexual. The greater similarity in sexual orientation in pairs of identical twins compared to non-identicals probably arises from their greater similarity in genetic make-up. Hence genes have a considerable effect on one's preference for one sex or another. These studies have concentrated on genes and male homosexuality. The situation regarding lesbianism and genes is less clear. Nevertheless, the work of Lynn Hall of the New York University School of Medicine and others provides evidence that genetic factors do influence sexual orientation in women.[7]

Identical twins do not always share the same sexual orientation, indicating that there are environmental influences on whether one adopts a heterosexual or homosexual lifestyle. The involvement of genetic and environmental influences leads us to consider the effect of the immune system on sexual orientation. One line of evidence which links the immune system with sexual orientation is the 'big brother effect'.

There are few things that are reliably associated with homosexuality. However, if you are male, one factor is known to influence your chances of being gay – the number of older brothers you have. We owe much of our knowledge of this surprising fact to the work of Ray Blanchard, a psychologist at the Centre for Addiction and Mental Health in Toronto. Blanchard's first study consisted of samples of 302 gay men and the same number of heterosexual men. He found that the gay men had an average of 1.3 older brothers compared to an average of 0.96 older brothers among the heterosexual men. The number of older sisters or younger brothers and sisters had no effect on the likelihood of being gay. This effect has been confirmed in Canada, the US and Europe, and Blanchard has calculated that the odds of being gay increase by about 33% with each older sibling.[8,9] This may appear to be surprisingly high, but given that the chance of being gay for men is only about one in fifty then it turns out that approximately one in seven gay men owe their sexual orientation to the influence of the big brother effect.

It is tempting to assume that if we can identify the causal factor for the big brother effect it may in some way show us why all gay men are gay. Blanchard himself focuses on immunity in his attempts to explain it, suggesting the influence of the mother's immune system on the developing child. Cells from the foetus can cross the placenta and even stay in the mother's blood for some time. Suppose the mother's immune system mounts an immune response to her developing son, but not daughter, and then 'remembers' when her next son is conceived and is developing. The maternal immune response might then affect her child's developing brain and change his sexual orientation. There is in fact some evidence for an unusual interaction between mother and son in homosexual males. Later-born children tend to be born heavier than first-borns. However, having an older brother tends to reduce this effect, so that a boy with two older brothers may be lighter at birth than his first-born brother. The suggestion of conflict between mother and developing foetus is supported by the observation that boys with older brothers have larger placentas than first-borns. The placenta is the route for food into the child from the mother. Do gay males suffer from a reduced food supply from their mother and develop a larger placenta as a response? Is this reduced food supply responsible for an effect on the centres for sexual orientation in the brain? We do not know. However, there is evidence that males who have older brothers appear to experience some prenatal interaction with their mother which first-born males do not. This interaction may involve the mother's immune system in some way.

Gay men and women often show abilities and characteristics which lie somewhere between the norms for heterosexual men and women. The work of Geoff Sanders of London Metropolitan University has shown that heterosexual men tend to score higher than heterosexual women in tests involving judgements of shape and other visual and spatial tasks, but gay men's scores more closely resemble those of heterosexual women than men.[10] A similar picture arises when we examine language skills. Heterosexual women and gay men have superior verbal abilities compared to heterosexual men. Visuo-spatial processing and linguistic skills may be influenced by foetal and adult sex hormones. It is there-fore tempting to speculate whether, in comparison to heterosexual men, gay men may have low testosterone and high oestrogen. The evidence is very mixed on this topic, if not downright confusing. Indeed, some studies point to the possibility that homosexual men may have more testosterone than heterosexual men. For example, testosterone is responsible for penis growth in the foetus and new-born infant and in boys entering puberty. A study by Anthony Bogaert drawing on multiple samples totalling about 4,200 men found evidence that gays have longer and thicker penises than heterosexual men.[11] A further indication that homosexual men have been exposed to high foetal testosterone comes from studies of left-handedness. Males are more often left-handed than females.[12,13] The sex difference is found in children, and left-hand preference is associated with masculine finger ratio.[14] This suggests an association between high foetal testosterone and left-handedness. An important recent study by Martin Lalumière, Ray Blanchard and Ken Zucker has shown that among men left-handedness is more common in gays than heterosexuals.[15] Thus we have evidence that many homosexual men are 'feminised' in some of their abilities, such as verbal fluency and visuo-spatial perception, but 'masculinised' in things such as penis length and left-handedness.

To date there have been four published studies concerning finger ratios and homosexuality, two in the UK and two in the US. First, the English data: with my colleague Simon Robinson, I obtained finger ratios from eighty-eight English men who described themselves as homosexual or bisexual.[16] These participants reported their sexual experiences and fantasies. The majority of the sample was exclusively or almost exclusively homosexual both in sexual partners and in reported fantasies. This finding was consistent with the literature on male homosexuality. Male bisexuals are rare. An inspection of the finger ratios of the gay men showed them to be quite 'Casanova-like', with averages of 0.97 and 0.96 for the right and left hands respectively. Comparisons were then made between the 'gay ratios' and controls obtained from samples of English men who were recruited without knowledge of their sexual orientation. This general population sample showed higher ratios, with an average of 0.98 for both hands. We came to what might seem a surprising conclusion – that the gay men in our sample had experienced higher prenatal testosterone than men recruited from the general population.

A second, London-based study came to similar conclusions of higher masculinisation in homosexuals, but this time the authors, Qazi Rahman and Glenn Wilson of King's College London, considered both gay men and lesbians.[17] Their sample comprised 240 participants made up of sixty subjects in each of the following groups: gay men, heterosexual men, lesbians and heterosexual women. They found lower finger ratios in gay men compared to heterosexual men, and in lesbians compared to heterosexual women. However, while in the UK it appears that gay men and women have more 'Casanova-type' finger ratios than heterosexual men and women, different results have been obtained in the US.

Marc Breedlove of the University of California was the first to publish a study concerning finger ratio and homosexuality.[18] A sample of 720 participants was recruited from gay street fairs in the San Francisco area. The sample comprised 277 homosexual and 108 heterosexual men, together with 164 homosexual and 146 heterosexual women. The findings were not surprising for women: lesbians had lower finger ratios than heterosexual women. However, for men there were no strong differences between the average finger ratios of gays and heterosexuals. It seems that in the San Francisco area homosexual and heterosexual men have been exposed to similar amounts of testosterone and oestrogen before birth.

Finally, a study of finger ratio and female homosexuality was conducted by Janel Tortorice of Rutgers University.[19,20] An interesting feature of this study was the division of lesbians into 'butch' and 'femme' categories (i.e. the tendency to adopt a 'masculine' or 'feminine' role in a lesbian relationship). Tortorice found that lesbians had lower ratios than heterosexual women, confirming the findings of Breedlove's group and of Rahman and Wilson. However, butch lesbians had more masculine ratios than femme lesbians, whose ratios were similar to those of heterosexual females. It appears that the difference in finger ratios between lesbians and heterosexual women is largely the result of the masculinised fingers of butch lesbians. This conclusion was supported by Breedlove's group in a study of lesbians from the Oakland Gay Pride Mardi Gras.[21] Butch and femme roles have been interpreted as social constructs, a reflection of male–female roles within heterosexual relationships, and therefore lacking any biological basis. Tortorice's work suggests that this is not correct. Butch lesbians appear to have been exposed to higher prenatal testosterone and lower prenatal oestrogen than femme lesbians and heterosexual women. The butch and femme roles may therefore be a social construct founded on the bedrock of a real biological difference which is determined very early in development. In this respect they are similar to many human sex differences.

So what are we to make of these differences in finger ratios between heterosexuals and homosexuals? Clearly in men the relationship between finger ratios and sexuality is complex. It seems that in England gays are on average more 'Casanova-like' in their finger ratios than heterosexual men. In California both groups have similar levels of testosterone before birth. The picture is made more complicated by recently published data by Dennis McFadden[22] and Richard Lippa[23] which show that in some US populations gay men have more feminised finger ratios than heterosexual men.

In order to make sense of this highly complex picture we need a study that draws from a very large international sample of heterosexual and homosexual participants. The recent BBC Internet Sex Survey recruited some 255,116 participants from nearly 200 countries. The lengths of the second and fourth fingers were self-measured by participants, who also indicated their sexual orientation as heterosexual (straight), homosexual (gay) or bisexual. With such a large sample we can be assured that any statistically significant effects are real. With regard to sexual orientation, the data from the Survey clearly show that white homosexual men have higher (more feminised) finger-ratios than heterosexual men. This finding holds true across many countries, including the UK and US suggesting many of the earlier studies may have been flawed by poor measuring technique. The position in other ethnic groups is less clear and we must wait for larger samples from black and Chinese men before it is clear whether male homosexuals are universally more feminised in their finger ratios than heterosexual men. For women the Survey did not differentiate between butch and femme lesbians so that we must wait for confirmation that the former are masculinised in their finger-ratios. However, at present the evidence does suggest that butch lesbians have had high prenatal testosterone.

Clearly homosexuality is a most puzzling behaviour. It appears to be influenced by genes, but gays (particularly gay men) have fewer children than heterosexuals. Why does natural selection not remove the genes for homosexuality from the human population? This problem is not unique to humans. Homosexuality is widespread in other species. Our answer to 'why is homosexuality relatively common?' must recognize this fact.

It seems unlikely that homosexuality is in some way 'adaptive'. That is it does not seem to provide some advantage which overcomes the massive evolutionary disadvantage of a reduction in family size. For example there is no evidence that homosexuals increase the number or wellbeing of their relatives in a way that heterosexuals do not. We are then left with the impression that the mechanism determining our sexual orientation is 'error prone'. Of course this then raises the question why has natural selection not reduced such errors to a vanishingly small number. Perhaps the answer lies in a consideration of the action of sex hormones on the brain of the developing foetus. A gene for low foetal testosterone will result in reproductive handicaps to male foetuses (which may include homosexuality) but reproductive advantages to female foetuses (which may include heterosexuality). A gene for high foetal testosterone will confer reproductive advantages on male foetuses (which may include heterosexuality) and reproductive handicaps on female foetuses (which may include homosexuality). Such genes are sexually antagonistic in their action and are difficult to remove by natural selection. This is because they vary from being advantageous to disadvantageous as they pass from one sex to the other. Thus homosexuality in men may result from genes which give reproductive advantages to females and disadvantages to males.

Our consideration of homosexuality brings to a close our voyage through human sexual selection and its links with prenatal hormones, disease predispositions, athleticism, male display and finger ratio. However, there is one last question, that of the effects of prenatal hormones on the human evolution, which a consideration of finger ratios may offer new insights. I consider the links between finger ratio and the origin of humans in the next chapter.

XI

Fingers, Schizophrenia and the Feminised Ape

It hardly seems possible that our finger ratios, with their insignificant sex difference in the index and ring finger, might be able to lead us into important insights into major human diseases, sexual orientation and difficult questions regarding skin colour. However, because the origin of sex differences before birth is such an important process, we can use its marker, the finger ratio, to illuminate another great puzzle. Our fingers can tell us something about the origin of our species.

Finger ratios, I have suggested, are a sort of 'living fossil', a record of part of our early development, or ontogeny. There is an axiom in biology that in some way development, and in particular early development, has within it clues to the evolution of our species, our phylogeny. Hence the saying 'Ontogeny repeats phylogeny'. In common with many of the great generalisations in biology, this axiom contains much truth but must be used with caution.

Our finger ratios are a testament to the proportions of early testosterone and oestrogen that set us on our individual life course, but this living fossil within our fingers also bears the traces of the event that led to the origin of our species. Finger ratios may be the 'smoking gun' that points to the simple genetic innovations that had the most profound of consequences, the origin of modern humans.

Modern humans arose in East Africa between one hundred and two hundred thousand years ago. Prior to this event our ancestors were very human-like, sufficiently so that they could be placed into our genus, *Homo*. They walked erect and their brains were large.

However, it is probable that they lacked one essential human trait – the ability of the right and left sides of the brain to function with some degree of independence. When mankind evolved this innovation it possessed the potential to acquire the three universal characteristics of modern humans: a tendency towards right-handedness, the ability to elaborate language and a susceptibility to schizophrenia. It should come as no surprise that the essence of humanity should include handedness and language. But why a debilitating psychosis like schizophrenia? Tim Crow of Oxford University's Department of Psychiatry, and the late David Horrobin of Laxdale Research in Stirling, have argued that in order to understand the origin of humanity we must first understand why schizophrenia exists, and why it is relatively common.[1,2]

Schizophrenia shows itself in 'positive' or 'negative' symptoms. Examples of the former include such things as hallucinations in which the sufferer may hear voices, an intense interest in religious matters, and a tendency towards paranoia and disorganisation of thought. Negative symptoms include an inability to experience joy and a tendency towards impulsive behaviour. Schizophrenia is often a very serious mental illness, and sufferers, particularly male sufferers, have reduced family sizes. It is this reduction in family size that must give us pause, because schizophrenia is widespread, relatively common, and influenced by genes.

Schizophrenia is found in all cultures. It is not a product of the relaxation of natural selection in Westernised societies. Both industrial and traditional societies have rates of schizophrenia that approximate to 1% of the population.[3] This is your lifetime chance of developing the disease. However, if one of your parents suffered from schizophrenia then the probability rises substantially. Is this increase in risk the result of socialisation by parents? To answer this we must consider adopted children from families with or without a schizophrenic parent. Children from the former are more likely to show evidence of schizophrenia, manic depression, alcoholism and criminality. However, creativity, musical ability and intense interest in religion also appear to be associated with the burden of psychosis.[4] We do not have to study adopted children to see such patterns of disordered mental function – consider, for example, detailed Icelandic studies of schizophrenia. The population of Iceland is relatively isolated from immigration and has been studied intensively by Karlsson.[5] Records of the last two hundred years have revealed familiar associations between on the one hand madness and creativity and on the other stable behaviour without unusual achievement. Here we have the first clues as to why natural selection has not removed the genes for psychosis which lead to schizophrenia and manic depression – such selection would also remove the basis of human creativity. But we probably need more than this to balance the strong loss of fitness or reduction in family size which schizophrenia brings. In order to be precise about the advantages and disadvantages of genes for schizophrenia we must identify the nature of the mutation or mutations which lead to our 'species-signature', i.e. handedness, language and schizophrenia.

What was the last great evolutionary innovation that took us from a species with limited intellectual horizons to our present abilities, from an animal that could fashion hand axes from flint and maybe weapons from wood, to a people which has elaborated religion, art and warfare? I would suggest that this biological innovation was the latest in a long line of human mutations that involved an increase in prenatal oestrogen and a reduction in the amount of, or sensitivity to, prenatal testosterone. We are not the 'naked ape', the 'bipedal ape', the 'throwing ape' or the 'hunting ape'. What defines our humanity is that our foetus is bathed in high oestrogen and low testosterone. We are the 'oestrogenised ape'.

Let us turn to our finger ratios to see evidence of our oestrogenisation. First, consider the tendency towards right-handedness. In common with us, chimpanzees have extraordinary control of the fingers. They pick up food, throw objects, scratch and groom their bodies and those of others, and even use leaves, twigs, branches and stones as tools for drinking and obtaining food. While engaged in such tasks they often have a preferred hand, and so individual chimps can be said to show handedness. Now try a little experiment. Go to your local zoo and note the patterns of handedness in a group of apes such as chimpanzees. If there are enough animals you will see that about half show some inclination to use their right hand in preference to their left, while the remainder will use their left in preference to their right. They are not as a species right-handed. Now turn and look at the humans who are looking at the apes. Note which hand they point with, hold their ice cream with, or scratch their heads with. It tends to be their right. This is a fundamental difference between ourselves and the rest of our primate relatives.

Sex is important in hand preference. Right-handedness is not just characteristic of the human species, it is most often found in human females. Consistently across human groups women have been observed to show right-handedness in about 93% of cases. Males often have rates of right-handedness that are closer to 91%, meaning that left-side preference is 27% more frequent in males than females.[6] The common-sense explanation for this sex difference lies in early oestrogenisation. High levels of oestrogen and low levels of testosterone before birth favour right-handedness. What do our fingers tell us about this?

There has been one study of finger ratios and hand preference, in a sample of 285 Jamaican children aged five to eleven.[7] The sample was selected from a rural population on the south side of the island. Hand speed was calculated for each hand in each child using a peg-board test, a simple procedure involving the one-handed task of moving pegs from one row of holes to another. In general people take less time to do this with their right hand than their left, i.e. we are as a species faster with our right hand. In this study right-hand advantage was more strongly found in girls than boys. As expected, finger ratios were of the female type in girls and the 'Casanova type' in boys. More importantly, within each sex the female-type ratios were associated with a very fast time in the right hand compared to the left. This suggests that high prenatal oestrogen is related to the development of right-handed preference in the individual, and to the evolution of right-handedness in our species. What, though, of language?

The ability to learn language appears early in children's development. They learn the vocabulary and rules of sentence construction for their particular language with ease and usually without formal instruction. Thus the brain is programmed so that children simply infer the rules of sentence construction appropriate for their particular language. In order to explain this and other features of language acquisition Noam Chomsky and his colleagues have argued that there is within the brain a 'deep structure' which is universal to all languages and is a fundamental aspect of the brain of modern humans.[8]

However, linguistic ability is not uniform across all humans. There is a sexual dimension, many aspects of language showing female superiority. Take verbal fluency. Suppose that we ask subjects to list as many words as possible starting with, say, the letters F, A and S. You will find considerable variation in recall between individuals, much of it related to sex. As one might expect, women have greater 'phonologically' based fluency than men. Try another test that relates not to sound but the meaning of words. Suppose we ask our subjects to recall words associated with groups such as musicians, animals, politicians, and sports people. This is some- times called the 'Varley' test.[9] Again you will find that this 'semantically' based fluency shows a female advantage. Hence those who have on average been exposed to high prenatal oestrogen, i.e. females, tend to excel at word fluency. Can we use the finger ratio to see evidence that within each sex that it is indeed high oestrogen and low testosterone that is driving this fluency? To date there has been one test of this idea. In a sample of one hundred women and one hundred men it was found that high 'semantically' based fluency was indeed related to female-type finger ratios. It seems that high prenatal oestrogen is associated with aspects of language in the individual and to the evolution of language in the human species.[10] This conclusion leads us to a consideration of schizophrenia.

The prevalence of schizophrenia does not reflect the frequency of character disorders and neurotic illness to be found in the general population. Among latent schizophrenics there are many psycho- paths who create stress for those around them, and many neurotics who keep their doctors busy without any reasonable chance of cure. Such tendencies as eccentricity, an inability to make social contact, an intense interest in religion, the conviction that thoughts are being inserted into one's head by others, hearing voices, suspicion and paranoia are common, and many people who display them are not considered mentally ill. Nevertheless, these 'schizotypal spectrum disorders' may precede the onset of schizophrenia proper, and they are common in the families of schizophrenics.[11] For example, adoption studies report rates of approximately 20% for schizotypal traits in children of parents with schizophrenia, but about 6% in children of parents of good mental health.[4] This means that many near-normal people are carrying the genes which when present in sufficient numbers, lead to schizophrenia.

Schizotypal tendencies can be measured by questionnaires. One such is the Oxford-Liverpool Inventory of Feelings and Experiences. The O-LIFE consists of 104 questions designed to reveal tendencies towards such things as magical thinking, attention difficulties, absence of friendships and impulsiveness. Thus the questions include: Have you ever felt that you have special, almost magical powers? Are you easily confused if too much happens at the same time? Are you much too independent to get involved with other people? Do you often have the urge to break or smash things? Magical thinking and attention difficulties are often grouped into 'positive' symptoms and absence of friendships and impulsiveness into 'negative' symptoms. A recent unpublished study by my group has considered the relationship between finger ratios and schizotypy in a sample of 460 subjects. It was found that female-type finger ratios are weakly but significantly associated with high O-LIFE scores for schizotypal tendencies. The relationship was strongest in women and for positive symptoms. This result suggests that schizotypy, and probably schizophrenia, are associated with high oestrogen exposure before birth. It also supports an evolutionary association between high oestrogen and the origin of schizophrenia.

Female-like finger ratios and prenatal oestrogen, handedness, language ability and schizotypy/schizophrenia are intertwined in the human-speciation event. However, these associations take us only part of the way to understanding schizophrenia. What is the nature of the mechanism that actually causes the disease? Janice Stevens of the Oregon Health Sciences University may have come close to an explanation.[12] Stevens points out that sex hormones and schizophrenia are closely related. For example, the onset of symptoms usually occurs during the reproductive period, with a marked peak between twenty and twenty-four years. During the onset of puberty and at the peak of fertility the brain is exposed to a flood of sex hormones, including oestrogen, testosterone and the hormones that stimulate their production. In general, reproductive hormones have what is known as a 'neuroexcitatory' effect. They stimulate certain parts of the brain. What stops the brain from being overwhelmed from the effect of sex hormones is the production of, or increased sensitivity to, inhibitors such as dopamine and serotonin. This means one must get this balance between excitation and inhibition right in order to maintain good mental health. Stevens points out that the most effective treatments for schizophrenia target brain receptors which cause inhibition, while schizophrenia may be caused by drugs such as LSD and amphetamines which actually enhance inhibition. Schizophrenics, then, may have too many inhibitors of oestrogen in the brain, and this may block the action of the hormone.

Suppose that the numbers of inhibitory brain receptors are determined early in the foetus as a response to oestrogen levels. High oestrogen stimulates more inhibitory receptors, while low oestrogen gives fewer inhibitory receptors. A foetus with high prenatal oestrogen will then have high levels of inhibitors that will help it deal with high adult levels of the hormone. But what if there is then a mismatch between foetal and adult levels of oestrogen? If the latter are too low the result may be schizophrenia. This logic suggests that treatment of schizophrenia, particularly in women, should involve both oestrogen and drugs which block receptors of inhibitors such as dopamine and serotonin.

Oestrogen is likely to be only one side of the story of the origin of our species. Humans are the most testosterone-insensitive of the primates, and this testosterone insensitivity is also likely to be an important evolutionary innovation. As we saw in Chapter III (pp. 32–9), the androgen receptor enables testosterone to trigger a cascade of effects as it 'switches on' some genes whose products then switch on others. We all have a gene (AR) located on our X chromosome which enables us to make this receptor. Males have one X chromosome and hence only one copy of the AR, inherited from their mother. Females have two X chromosomes and therefore have copies of two ARs, one from their mother and one from their father, though one X chromosome is deactivated early in development so that they too have only one active AR.

The AR contains a repetitive region that determines the number of glutamine units present in one region of the receptor molecule: as we have seen, this number determines sensitivity to testosterone. Let us compare our species to other mammalian groups in this regard. The AR in mice and rats,[13] for example, shows no variation from individual to individual. All members of these species have only one glutamine unit in their androgen receptor, indicating high testosterone sensitivity. Within our own order, the primates, our closest relatives are the apes. These include the tree-dwelling gibbons and the larger, more human-like orangutans, gorillas and chimpanzees. Philippe Djian and his colleagues at the Hammersmith Hospital, London, have shown that apes display the beginnings of variation in the number of glutamines present in the androgen receptor molecule, and evidence of an increase in testosterone insensitivity.[14] While gibbons have four glutamines and no variation from individual to individual, gorillas have six to seventeen glutamines and chimpanzees eight to fourteen. Compared to the human average of twenty-one glutamine units and a range of eleven to thirty-one in the androgen receptor, these numbers show evidence of increasing testosterone insensitivity as we move from small apes through human-like apes to humans. It is tempting to suggest that high human insensitivity to testosterone was acquired at the human speciation-event along with reductions in prenatal testosterone and increases in prenatal oestrogen. There is as yet no evidence that high numbers of glutamine units in the androgen receptor are associated with right-handedness, verbal fluency and schizotypy. However, given that our finger ratios are related to the structure of our androgen receptor gene, I think that when we look for these associations we will find them.

Given the relationships between fingers and the androgen receptor[15] it is not surprising that low glutamine levels in the receptor and masculine finger ratios are both related to high sperm counts and large family size in men, whereas in women they are associated with lowered fertility and small family size. High numbers of glutamine units in the androgen receptor and female-type finger ratios are associated with low male fertility but increased female fertility. In Chapter VI (pp. 71–80) I argued that men in societies with intense male competition for wives will show high testosterone, masculine finger ratios, low resistance to disease and dark skin. There is now evidence of ethnic differences in the andro-gen receptor that support an association between high testosterone sensitivity and black skin. A study conducted by Louisiana State University has shown that a sample of sixty-five African-American men had an average of nineteen glutamine units in their androgen receptor. In contrast, the average number of glutamines for 130 white Americans was twenty-one. The overall sample showed that black males were twice as likely as white to have fewer than twenty glutamine units, hence were more sensitive to testosterone than the white American sample.[16] This work is at a preliminary stage, and it is important to realise that we are not concerned here with racial differences. These are male-to-male competition differences that cut across simple black/white/oriental categories. However, I believe that an overall association between intense male competition for wives, masculine digit ratios, high sensitivity to testosterone and black skin will emerge.

We are now in a position to discuss the origin of our species in relation to early sex hormones and finger ratios. My belief is that since splitting from the gorilla–chimp–human line, human evolu-tion has been driven by successive mutations that have reduced prenatal testosterone and sensitivity to testosterone, and increased our prenatal oestrogen. We have moved away from a 'Casanova-like' ratio towards a more female-like ratio.

Compare humans and our closest living relatives, the great apes (orangutans, gorillas and chimpanzees). In general we are more slender or gracile in our physical features: our brow ridges are reduced, our jaw is small and does not protrude forwards, thus giving a flat face, our teeth are reduced, our skull is large and vaulted and lacks a prominent ridge for muscle attachment on its upper surface, and the skeletal elements of our limbs are slender. It is often said that in comparison to apes we appear neotenous or child-like, and this has led some to argue that we are in fact a species of neotenous ape, i.e. an animal that has retained many of its juvenile features and has reached sexual maturity while still retaining these features. I would like to suggest an alternative to the human–child–ape comparison. We are not a child-like ape but a feminised or oestrogenised ape. Our relative lack of brow ridges, small jaw and slender skeleton are all female rather than male characteristics. Compared to women, human males have thicker brow ridges, larger jaws and a more robust skeleton. However, in these traits they are far exceeded by male and female great apes. There is some evidence that apes have very masculinised finger ratios compared to humans, but we do not yet know what the data tells us regarding ape and human finger ratios. However, this does seem consistent with the human trend of feminised features. One exception to this observation is the gibbon. Gibbons are the smallest of the apes, are almost entirely tree-dwelling, and live in family groups which are based on monogamy. Male and female gibbons are very similar in size and body-type, presumably because male-to-male competition for females is low. I would therefore expect gibbons and humans to have similar finger ratios.

The trend for the human skeleton to become less robust since humans and great apes separated between 5 and 10 million years ago is consistent with a tendency towards increasing oestrogenisation. Early in this evolutionary line are the Australopithecines, small-brained, ape-like animals that walked erect. *Australopithecus afarensis* and *Australopithecus africanus* were found in Africa between 3.9 and 2 million years ago. *A. afarensis* had ape-like features of the skull, including thick brow ridges, protruding jaws and large molar teeth, but the canine teeth were reduced compared to chimpanzees. There is evidence that *A. afarensis* differed markedly in size between males and females, an indication of strong male competition for females. It is to be expected that finger ratios would be masculinised in this hominid. The reduction in canine teeth continued in *A. africanus*, although the molar teeth were slightly larger. *A. afarensis* and *A. africanus* and related species are known as the gracile or slender Australopithecines, as opposed to such species as *A. aethiopithecus*, *A. robustus* and *A. boisei*, the robust Australopithecines. These last had heavy skulls and large teeth. The gracile rather than the robust species lie close to the human line, although they were markedly more robust than modern humans.

The line to modern humans continues with the African species *Homo habilis*, the first hominid to show evidence of tool use, and then to the geographically widespread *Homo erectus*. Both *H. habilis* and *H. erectus* continue the gracile trend, although the skeleton is still more robust than that of modern humans. The archaic forms of our species, *Homo sapiens*, appear about 500,000 years ago. The skeleton, teeth and skull of *Homo sapiens* (*archaic*) is less robust than *H. erectus* but more robust than modern humans. Modern *Homo sapiens* appeared some 120,000 years ago. The skeleton is very gracile, the jaws are small and the eyebrow ridges much reduced or absent.

The long-term trend of reduction in the robust nature of the skull does not end here, however: humans of about thirty thousand years ago were about 30% more robust in the face, teeth and jaws than present-day forms, and those of about ten thousand years ago were 10% to 15% more robust. Today humans are very gracile in comparison to our ancestors. There is some variation in traits such as tooth size, which are related to long-term differences in food types, but overall our skeleton bears the marks of low prenatal testosterone and high prenatal oestrogen.

An increase in prenatal oestrogen and a reduction in prenatal testosterone are likely to have a number of harmful consequences. Most of these negative consequences are likely to impact on males. A loss of strength in the skeleton and muscles, a deterioration in the efficiency of the heart and the blood vessels, and a reduction in sperm numbers and quality are all to be expected. Why should the human line embrace repeated mutations which lead to the decline of male competitiveness and fertility? We are less robust than orangutans, gorillas and chimps, but we are cleverer. The conclusion seems obvious: progressive oestrogenisation has given us more dexterity, enabling us to become tool-makers, and made it possible for us to elaborate language. In short, an oestrogenised ape is a smart ape.

It seems appropriate to finish with an overview of modern humans with regard to our present levels of prenatal sex hormones. As a species we are feminised in comparison to our closest living relatives and to our ancestors. However, within the confines of our species our finger ratios indicate marked geographical variation in the amount of early exposure to oestrogen and testosterone. Figure 11.1 gives average finger ratios for ten human populations, many of which we have seen in the course of this book. I have divided them into four groupings – Europe (Poland, Spain, England, Hungary and Germany), India (Yanadi and Sugali tribal groups, and Hungarian Gypsies, who are derived from Indian tribes), East Asia (Han Chinese from Beijing and Japanese from Tokyo) and black (Zulus and Afro-Caribbean Jamaicans).

Female-like finger ratios are found in European populations, but there is considerable variation: Polish, Spanish and English peoples are more oestrogenised than Germans. The Indian and East Asian groups have intermediate ratios, and African peoples are more testosteronised before birth than most populations. Throughout these marked differences in finger ratios we find consistent sex differences, although they are weaker than the observed ethnic variation.

As mentioned in Chapter VII (pp. 81–94), I believe that a consideration of ancient marriage systems is the key to understanding some of this ethnic variation.[17] Monogamy is the most common European marriage system and it is associated with low male-to-male competition for women: sure enough, European peoples

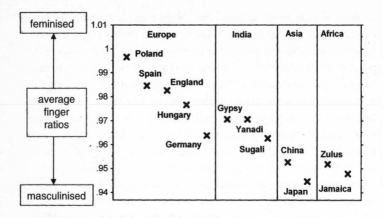

FIGURE 11.1

An overall view of average finger ratios in four areas of the world and in various homosexual populations. With some exceptions, Europeans tend to have female-type finger ratios. The ratios of Indians and Gypsies are of intermediate value, and East Asian, Afro-Caribbean and African populations have 'Casanova-like' ratios.

appear to be the most oestrogenised of our species, with average finger ratio values of 1.00 down to 0.96.

Peoples in the Indian subcontinent and in Eastern Asia show a mix of monogamy and occasional polygyny. Here our data show evidence for intermediate levels of oestrogenisation, with average finger ratios varying from 0.97 down to 0.95. The East Asian samples show masculinised ratios in Chinese and Japanese populations. Monogamy is common in the Shantung Chinese of the north, but occasional polygyny is found in the Min Chinese of the south. The Japanese sample is perhaps the most surprising, with an average ratio of 0.95. This is low for a monogamous people. Nevertheless, there is evidence for high prenatal testosterone and low prenatal oestrogen in both the Chinese and Japanese populations, which enjoy low rates of heart disease and breast cancer.

We find low average ratios of around 0.95 in the Zulus of South Africa and in Afro-Caribbean Jamaicans. In common with almost all sub-Saharan peoples, Zulus are polygynous, and, as I have discussed, high prenatal testosterone is favoured in polygynous societies. It is in the sub-Saharan region that many indigenous peoples are found with a skin colour that is darker than expected from a consideration of UV light intensities alone. I have argued that black skin is associated with polygyny, low latitudes and low finger ratios. Afro-Caribbean people in Jamaica have very low finger ratios, and black football players in the UK have lower finger ratios than white.[10] I think it likely that indigenous and immigrant groups of black people will typically show low finger ratios, a theory supported by data from variation in the androgen receptor gene. African-American men have lower numbers of glutamine repeats in their receptor, i.e. are more sensitive to testosterone, than white American men. With high prenatal testosterone and high sensitivity to testosterone, black males are likely to have efficient cardio-vascular systems, high running speeds, well-developed athleticism and high sperm counts. However, set against these advantages testosterone may have a harmful effect on the developing immune system. Thus polygynous black populations may be prone to lowered immunity and high disease loads caused by bacteria, fungi and parasites.

I conclude that our fingers provide clues as to what has happened to humans since we split from our ape ancestors some 5 million or so years ago. Comparisons between modern humans, apes, and the series of early hominids that led to modern humans suggest that we are at the end of a long line of events that have resulted in the progressive feminisation of our species. This process has increased our manual dexterity, made language possible and enabled us to become more intelligent. However, it has brought with it a reduction in male competitiveness, with an increased tendency for a loss in cardiovascular efficiency and a reduction in sperm numbers and viability. In short, we are the feminised ape, and we are intelligent. Yet this intelligence is continuing to evolve in a way that might have serious consequences for the long-term success of our species.

References

A Tale of Two Fingers

1. Collaer, M. L., and Hines, M., 1995. 'Human Behavioral Sex Differences: a Role for Gonadal Hormones during Development?' *Psychological Bulletin*, 118: 55–107.
2. Brabin, L., and Brabin, B. J., 1992. 'Parasitic Infections in Women and Their Consequences'. *Advances in Parasitology*, 31: 1–81.
3. Hepper, P. G., Shannon, E. A., and Dornan, J. C., 1997. 'Sex Differences in Foetal Mouth Movements'. *Lancet*, 350: 1820.
4. Casanova, G., 1984. *History of My Life*, Vol. 11, translated into English in accordance with the original French manuscript by Willard R. Trask. London: Longman.
5. Roettgen, S., 1993. *Anton Raphael Mengs 1728–1779 and His British Patrons*. London: English Heritage.
6. Gray, H., 1858. *Gray's Anatomy*. Bristol: Paragon Press.
7. Wessels, H., Lue, T. F., and McAninch, J. W., 1996. 'Penile Length in the Flaccid and Erect State: Guidelines for Penile Augmentation'. *Journal of Urology*, 156: 995–7.
8. Bain, J., and Siminowski, K., 1993. 'The Relationship among Height, Penile Length, and Foot Size'. *Annals of Sex Research*, 6: 231–5.
9. Bogaert, A. F., and Herschberger, S., 1999. 'The Relation between Sexual Orientation and Penile Size'. *Archives of Sexual Behavior*, 28: 213-21.
10. Spyropoulos, E., Borousas, D., Mavrikos, S., Dellis, A., Bourounis, M., and Athanasiadis, S., 2002. 'Size of External Genital Organs and Somatometric Parameters among Physically Normal Men Younger than 40 Years Old'. *Urology*, 60: 485–9.
11. Voracek, M., and Manning, J. T., 2003. 'Length of Fingers and Penis are Related through Foetal *Hox* Gene Expression'. *Urology*, 62, 201.
12. Kondo, T., Zakany, J., Innis, J., and Duboule, D., 1997. 'Of Fingers, Toes and Penises'. *Nature* 390: 29.

13. Ecker, A., 1875. 'Einige Bemerkungen über einen Schwankenden Charakter in der Hand des Menschen'. *Archiv für Anthrop* (*Braunschweig*), 8: 67–75.

14. Schultz, A. H., 1924. 'Growth Studies on Primates bearing upon Man's Evolution'. *American Journal of Physical Anthropology* 7: 149–64.

15. Sorell, W., 1968. *The Story of the Human Hand.* London: Weidenfeld and Nicholson.

16. Darwin, C., 1871. *The Descent of Man, and Selection in Relation to Sex.* London: Raven Press.

17. Baker, F., 1888. 'Anthropological Notes on the Human Hand'. *American Anthropologist* 1: 51–76.

18. Phelps, V. R., 1952. 'Relative Index Finger Length as a Sex-Influenced Trait in Man'. *American Journal of Human Genetics* 4: 72–89.

Fingers, Sex, Class and Ethnicity

1. Manning, J. T., Scutt, D., Wilson, J., and Lewis-Jones, D. I., 1998. 'The Ratio of 2nd to 4th Digit Length: a Predictor of Sperm Numbers and Levels of Testosterone, LH and Oestrogen'. *Human Reproduction* 13: 3000–4.

2. Manning, J. T., and Bundred, P., 2000. 'The Ratio of 2nd to 4th Digit Length: a New Predictor of Disease Predisposition?' *Medical Hypotheses* 54: 855–7.

3. Sluming, V. A., and Manning, J. T., 2000. 'Second to Fourth Digit Ratio in Elite Musicians: Evidence for Musical Ability as an Honest Signal of Male Fitness'. *Evolution and Human Behavior* 21: 1–9.

4. Manning, J. T., and Taylor, R. P., 2001. '2nd to 4th Digit Ratio and Male Ability in Sport: Implications for Sexual Selection in Humans'. *Evolution and Human Behavior* 22: 61–69.

5. Manning, J. T., and Leinster, S., 2001. 'The Ratio of 2nd to 4th Digit Length and Age at Presentation of Breast Cancer: a Link with Prenatal Oestrogen'. *The Breast* 4: 355–7.

6. Williams, T. J., Pepitone, M. E., Christensen, S. E., Cooke, B. M., Huberman, A. D., Breedlove, N. J., Breedlove, T. J., Jordan, C. L., and Breedlove, S. M., 2000. 'Finger Length Patterns Indicate an Influence of Foetal Androgens on Human Sexual Orientation'. *Nature* 404: 455.

7. Manning, J. T., Barley, L., Lewis-Jones, I., Walton, J., Trivers, R. L., Thornhill, R., Singh, D., Rhode, P., Bereckzei, T., Henzi, P., Soler, M., and Sved, A. 2000. 'The 2nd to 4th Digit Ratio, Sexual Dimorphism, Population Differences and Reproductive Success: Evidence for Sexually Antagonistic Genes'. *Evolution and Human Behavior* 21: 163–83.

8. Barker, D.

9. Migeon, C. J., and Wisniewski, A. B., 1998. 'Review – Sexual Differentiation: from Genes to Gender'. *Hormone Research* 50: 245–51.

10. Garn, S. M., Burdi, A. R., Babler, W. J., and Stinson, S., 1975. 'Early Prenatal Attainment of Adult Metacarpal-Phalangeal Rankings and Proportions'. *American Journal of Physical Anthropolology* 43: 327–32.

11. Danforth, C. H., 1921. 'Distribution of Hair on the Digits in Man'. *American Journal of Physical Anthropology* 4: 189–204.

12. Garn, S. M., 1951. 'The Use of Middle Phalangeal Hair in Population Studies'. *American Journal of Physical Anthropology* 9: 325–33.

13. Gray, H., 1858. *Gray's Anatomy.* Bristol: Paragon Press.

14. Winkler, E. M., and Christiansen, K., 1993. 'Sex Hormone Levels and Body Hair Growth in !Kung San and Kavango Men from Namibia'. *American Journal of Physical Anthropology* 92: 155–64.

15. Holt, S. B., 1968. *The Genetics of Dermal Ridges.* Springfield: Charles C. Thomas.

16. Jamison, C. S., Meier, R. J., and Campbell, B. C., 1993. 'Dermatoglyphic Asymmetry and Testosterone Levels in Normal Males'. *American Journal of Physical Anthropology* 90: 185–98.

17. Kondo, T., Zakany, J., Innis, J., and Duboule, D., 1997. 'Of Fingers, Toes and Penises'. *Nature* 390: 29.

18. Coates, M., 1996. 'The Devonian Tetrapod *Acanthostega Gunnari Jarvik*: Postcranial Anatomy, Basal Tetrapod Interrelationships and Patterns of Skeletal Evolution'. *Transactions of the Royal Society of Edinburgh – Earth Sciences* 87: 363–72.

19. Runciman, W. G., 1998. *The Social Animal.* London: HarperCollins.

20. Herrnstein, R. J., and Murray, C., 1994. *The Bell Curve: Intelligence and Class Structure in American Life.* New York: Simon and Schuster.

21. Phillips, G. B., Pinkernell, B. H., and Jing, T. Y., 1994. 'The Association of Hypotestosteronemia with Coronary Artery Disease in Men'. *Arteriosclerosis and Thrombosis* 14: 701–6.

22. Entrican, J. H., Beach, C., Carroll, D., Klopper, A., Mackie, M., and Douglas, A. S., 1978. 'Raised Plasma Oestradiol and Oestrone Levels in Young Survivors of Myocardial Infarction'. *Lancet* 2: 487–90.

23. Aksut, S. V., Aksut, G., Karamehmetoglu, A., and Oram, E., 1986. 'The Determination of Serum Estradiol, Testosterone and Progesterone in Acute Myocardial Infarction'. *Japanese Heart Journal* 27: 825–37.

24. Dabbs, J. M., Jr, 1992. 'Testosterone and Occupational Achievement'. *Social Forces,* 70: 813–24.

25. Manning, J. T., 2000. *Digit Ratio: A Pointer to Fertility, Behaviour and Health.* New Brunswick: Rutgers University Press.

26. Ellis, L., and Nyborg, H., 1992. 'Racial/Ethnic Variations in Male Testosterone Levels: a Probable Contributor to Group Differences in Health'. *Steroids* 57: 72–5.

Fingers and Sex Hormones

1. Migeon, C. J., and Wisniewski, A. B., 1998. 'Review – Sexual Differentiation: from Genes to Gender'. *Hormone Research* 50: 245–51.
2. Manning, J. T., 2002. *Digit Ratio: a Pointer to Fertility, Behavior and Health.* New Brunswick: Rutgers University Press.
3. Raine, A., 1993. *Psychopathology of Crime.* San Diego: Academic Press.
4. Evans, D. J., Hoffmann, R. G., Kalkhoff, R. K., and Kissebah, A. H., 1983. 'Relationship of Androgenic Activity to Body Fat Topography, Fat Cell Morphology, and Metabolic Aberrations in Premenopausal Women'. *Journal of Clinical Endocrinology and Metabolism* 57: 304–10.
5. Singh, D., 1993. 'Adaptive Significance of Female Physical Attractiveness: Role of Waist-to-Hip Ratio'. *Journal of Personality and Social Psychology* 65: 293–307.
6. Singh, D., and Young, R. K., 1995. 'Body Weight, Waist-to-Hip Ratio, Breasts and Hips: Role in Judgements of Female Attractiveness and Desirability for Relationships'. *Ethology and Sociobiology* 16: 483–507.
7. Manning, J. T., Trivers, R. L., Singh, D., and Thornhill, R., 1999. 'The Mystery of Female Beauty'. *Nature* 399: 214–15.
8. Okten, A., Kalyoncu, M., and Yaris, N., 2002. *Early Human Development* 70: 47–54.
9. Brown, W. M., Hines, M., Fane, B. F., and Breedlove, S. M., 2002. 'Masculinised Finger Length Patterns in Human Males and Females with Congenital Adrenal Hyperplasia'. *Hormones and Behavior* 42: 380–6.
10. Choong, C. S., Kemppainen, J. A., Zhou, Z. X., and Wilson, E. M., 1996. *Molecular Endocrinology* 10: 1527–35.
11. Krithivas, K., Yurgalevitch, S. M., Mohr, B. A., Wilcox, C. J., Batter, S. J., Brown, M., Longcope, C., McKinlay, J. B., and Kantoff, P. W., 1999. 'Evidence that the CAG Repeat in the Androgen Receptor Gene is Associated with the Age-Related Decline in Serum Androgen Levels in Men'. *Journal of Endocrinology* 162: 137–2.
12. Sartor, O., Zheng, Q., and Eastham, J. A., 1999. 'Androgen Receptor Genes CAG Repeat Length Varies in a Race-Specific Fashion in Men without Cancer'. *Urology* 53: 378–80.
13. Stanford, J. L., Just, J. J., Gibbs, M., Wicklund, K. G., Neal, C. L., and Blumstin, B. A., 2000. 'Polymorphic Repeats in the Androgen Receptor

Gene: Molecular Markers of Prostate Cancer Risk'. *Cancer Research* 57: 1194–8.

14. Manning, J. T., Bundred, P. E., and Flanagan, B. F., 2002. 'The Ratio of 2nd to 4th Digit Length: a Proxy for Transactivation Activity of the Androgen Receptor Gene?' *Medical Hypotheses* 59: 334–6.

15. Von Eckardstein, S., Syska, A., Gromol, J., Kamischke, A., Simoni, M., and Nieschlag, E., 2001. 'Inverse correlation Between Sperm Concentration and Number of Androgen Receptor CAG Repeats in Normal Men.' *Journal of Clinical Endocrinology and Metabolism* 86: 2585–90.

16. Mifsud, A., Ramirez, S., and Yong, E. L., 2000. 'Androgen Receptor Gene CAG Trinucleotide Repeats in Anovulatory Infertility and Polycystic Ovaries'. *Journal of Clinical Endocrinology and Metabolism* 85: 3484–8.

17. Comings, D. E., Chen, C., Wu, S., and Muhleman, D., 1999. 'Association of the Androgen Receptor Gene (AR) with ADHD and Conduct Disorder'. *Neuroreport* 14: 1589–92.

18. Manning, J. T., Barley, L., Lewis-Jones, I., Walton, J., Trivers, R. L., Thornhill, R., Singh, D., Rhode, P., Bereckzei, T., Henzi, P., Soler, M., and Sved, A., 2000. 'The 2nd to 4th Digit Ratio, Sexual Dimorphism, Population Differences and Reproductive Success: Evidence for Sexually Antagonistic Genes'. *Evolution and Human Behavior* 21: 163–83.

19. Manning, J. T., Henzi, P., Venkatramana, P., Martin, S., and Singh, D., 2003. '2nd to 4th Digit Ratio: Ethnic Differences and Family Size in English, Indian and South African Populations'. *Annals of Human Biology* 30: 579–88.

20. Palermo, G., Joris, H., Devroey, P., and Van Steirteghem, A. C., 1992. 'Pregnancies after Intracytoplasmic Injection of Single Spermatozoon into an Oocyte'. *Lancet* 340: 17–18.

21. Sutcliffe, A. G., 2002. *IVF Children: the First Generation*. London: Parthenon.

22. Wood, S. J., Vang, E., Manning, J. T., Walton, J., Troup, S., Kingsland, C. R., and Lewis-Jones, I. D., 2003. 'The ratio of 2nd to 4th Digit Length in Azoospermic Males Undergoing Surgical Sperm Retrieval: Predictive Value for Sperm Retrieval and on Subsequent Fertilisation and Pregnancy Rates in IVF/ICSI Cycles'. *Journal of Andrology* 24: 871–7.

23. Rvdokimova, V. N., and Nazaarenko, S. A., 2001. 'Genetic Variability for Number of CAG-Repeats in the X-Linked Androgen Receptor Gene and Embryonal Cell Death in Humans'. *Genetika* 37: 248–55.

24. Manning, J. T., Martin, S., Trivers, R. L., and Soler, M., 2002. '2nd to 4th Digit Ratio and Offspring Sex Ratio'. *Journal of Theoretical Biology* 217: 93–5.

25. Rebback, T. R., Kantoff, P. W., Krithivas, K., Neuhausen, S., Blackwood, M. A., and Godwin, A. K., 1999. 'Modification of BRCA1-Associated Breast Cancer Risk by the Polymorphic Androgen-Receptor CAG Repeat'. *American Journal of Human Genetics* 64: 1371–7.

26. Manning, J. T., and Leinster, S., 2001. '2nd to 4th Digit Ratio and Age at Presentation of Breast Cancer'. *The Breast* 10: 355–7.

27. Manning, J. T., Bundred, P. E., Newton, D. J., and Flanagan, B. F., 2003. 'The 2nd to 4th Digit Ratio and Variation in the Androgen Receptor Gene'. *Evolution and Human Behavior* 30: 579–88.

Fingers and Personality

1. Buss, D., 1995. 'Psychological Sex Differences: Origins through Sexual Selection'. *American Psychologist* 50: 164–8.

2. Baron-Cohen, S., and Hammer, J., 1997. 'Is Autism an Extreme Form of the Male Brain?' *Advances in Infancy Research* 11: 193–217.

3. Collaer, M. L., and Hines, M., 1995. 'Human Behavioral Sex Differences: a Role for Gonadal Hormones during Early Development?' *Psychological Bulletin* 118: 55–107.

4. Williams, J. H. G., Greenhalgh, K. D., and Manning, J. T., 2003. 'Second to Fourth Finger Ratio and Possible Precursors of Developmental Psychopathology in Pre-School Children'. *Early Human Development* 72: 57–65.

5. Simonoff, E., Pickles, A., Meyer, J. M., Silberg, J. L., Maes, H. H., and Loeber, R., 1997. 'The Virginia Twin Study of Adolescent Behavioural Development. Influences of Age, Sex, and Impairment on Rates of Disorder'. *Archives of General Psychiatry* 54: 801–8.

6. McCrae, R. R., and Costa, P. T., Jr, 1997. 'Personality Trait Structure as a Human Universal'. *American Psychologist* 52: 509–16.

7. Budaev, R. R., 1999. 'Sex Differences in the Big Five Personality Factors: Testing an Evolutionary Hypothesis'. *Personality and Individual Differences* 26: 801–13.

8. Fink, B., Manning, J. T., and Neave, N., 2004. 'Second to Fourth Digit Ratio and the "Big Five" Personality Factors'. *Personality and Individual Differences* 37: 495–503.

9. Austin, E. J., Manning, J. T., McInroy, K., and Mathews E., 2002. 'An Investigation of the Associations between Personality, Cognitive Ability and Digit Ratio'. *Personality and Individual Differences* 33: 1115–24.

10. Wilson, G. D., 1983. 'Finger Length as an Index of Assertiveness in Women'. *Personality and Individual Differences* 4: 111–12.

11. Bem, S. L., 1981. *Bem Sex Role Inventory: Professional Manual.* Consulting Psychological Press.

12. Csatho, A., Osvath, A., Bicsak, E., Karadi, K., Manning, J. T., and Kallai, J., 2003. 'Sex Role Identity Related to the Ratio of Second to Fourth Digit Length in Women'. *Biological Psychology* 62: 147–56.

13. Gladue, B. A., 1991. 'Aggressive Behavioral Characteristics, Hormones, and Sexual Orientation in Men and Women'. *Aggressive Behavior* 17: 313–26.

14. Dabbs, J. M., Jurkovic, G. L., and Frady, R. L., 1995. 'Saliva Testosterone and Cortisol among Late Adolescent Juvenile Offenders'. *Journal of Abnormal Child Psychology* 19: 469–78.

15. Banks, T., and Dabbs, J. M., 1996. 'Salivary Testosterone and Cortisol in a Delinquent and Violent Urban Subculture'. *Journal of Social Psychology* 136: 49–56.

16. Archer, J., 1991. 'The Influence of Testosterone on Human Aggression'. *British Journal of Psychology* 82: 1–28.

17. Archer, J., 1994. 'Testosterone and Aggression'. *Journal of Offender Rehabilitation* 21: 3–25.

18. Mittwoch, U., and Mahadevaiah, S., 1980. 'Additional Growth: a Link between Mammalian Testes, Avian Ovaries, Gonadal Asymmetry in Hermaphrodites and the Expression of H-Y Antigen'. *Growth* 44: 287–300.

19. Gray, H., 1858. *Gray's Anatomy.* Bristol: Paragon Press.

20. Trivers, R. L., Manning, J. T., Thornhill, R., and Singh, D., 1999. 'The Jamaican Asymmetry Project: a Long-Term Study of Fluctuating Asymmetry in Rural Jamaican Children'. *Human Biology* 71: 417–30.

21. Baron-Cohen, S., and Bolton, P., 1993. *Autism: the Facts.* Oxford: Oxford University Press.

22. Baron-Cohen, S., 2002. 'The Extreme Male Brain Theory of Autism'. *Trends in Cognitive Sciences* 6: 248–54.

23. Baron-Cohen, S., Wheelwright, S., Stott, C., Bolton, P., and Goodyer, I., 1997. 'Is There a Link between Engineering and Autism?' *Autism* 1: 101–9.

24. Marazziti, D., and Cassano, G. B., 2003. 'The Neurobiology of Attraction'. *Journal of Endocrinology Investigations* 26: 58–60.

25. Young, L. J., 2001. 'Oxytocin and Vasopressin as Candidate Genes for Psychiatric Disorders: Lessons from Animal Models'. *American Journal of Medical Genetics* 105: 53–4.

26. Hollander, E., Novotny, S., Hanratty, M., Yaffe, R., De Carria, C. M., Aronovitz, B. R., and Mosovich, S., 2003. 'Oxytocin Infusion Reduces Repetitive Behaviour in Adults with Autistic and Aspergers Disorders'. *Neuropsychopharmacology* 28: 193–8.

27. The Autism Research Unit, University of Sunderland. http://osirus.sunderland.ac.uk/autism/index.html

28. Charman, T., 2002. 'The Prevalence of Autism Spectrum Disorders. Recent Evidence and Future Challenges'. *European Child and Adolescent Psychiatry* 11: 249–56.

29. Baron-Cohen, S., 2003. *The Essential Difference: Men, Women and the Extreme Male Brain.* London: Allen Lane.

30. Wakefield, A. J., 1999. 'MMR Vaccination and Autism'. *Lancet* 354: 949–50.

31. Kaye, J. A., Melero-Montes, M. del Mar, and Jick, H., 2001. 'Mumps, Measles, and Rubella Vaccine and the Incidence of Autism Recorded by General Practitioners: a Time Trend Analysis'. *BMJ* 322: 460–3.

32. Manning, J. T., Baron-Cohen, S., Wheelwright, S., and Sanders, G., 2001. 'The 2nd to 4th Digit Ratio and Autism'. *Developmental Medicine and Child Neurology* 43: 160–4.

33. Wender, P. H., 1987. *The Hyperactive Child, Adolescent and Adult: Attention Deficit Disorder through the Lifespan.* Oxford: Oxford University Press.

34. Selikowitz, M., 1998. *Dyslexia and Other Learning Difficulties: the Facts.* Oxford: Oxford Medical Publications.

Fingers, Heart Attacks and Cancer of the Breast and Ovary

1. Wexler, L. A., 1999. 'Studies of Acute Coronary Syndromes in Women: Lessons for Everyone'. *New England Journal of Medicine* 341: 275–6.

2. Kmietowicz, Z., 1999. 'Heart Disease Mortality Declining with Fewer and Less-Deadly Attacks'. *BMJ* 318: 1307.

3. Rosano, G. M. C., 2000. 'Androgens and Coronary Heart Disease: a Sex Specific Effect of Sex Hormones?' *European Heart Journal* 21: 868–71.

4. Manning, J. T., and Bundred, P. E., 2000. 'The Ratio of 2nd to 4th Digit Length: a New Predictor of Disease Predisposition?' *Medical Hypotheses* 54: 855–7.

5. Levy, E. P., Cohen, A., and Fraser, F. C., 1973. 'Hormone Treatment during Pregnancy and Congenital Heart Attacks'. *Lancet* 1: 611.

6. Nora, J. J., Nora, A. H., Perinchief, A. G., Ingram, J. W., Fountain, A. K., and Peterson, M. J., 1976. 'Congenital Abnormalities and First-Trimester Exposure to Progestogen/Oestrogen'. *Lancet* 1: 313–14.

7. Manning, J. T., and Bundred, P. E., 2001. 'The Ratio of Second to Fourth Digit Length and Age at First Myocardial Infarction in Men: a Link with Testosterone?' *British Journal of Cardiology* 8: 720–3.

8. Kannel, W. B., Wolf, P. A., Castelli, W. P. and D'Agostino, R. B., 1987.

'Fibrinogen and Risk of Heart Disease. The Framingham Study'. *JAMA* 258: 1183–6.

9. Ernst, E. 1990. 'Plasma Fibrinogen – an Independent Cardiovascular Risk Factor'. *Journal of Internal Medicine* 227: 365–72.

10. Lip, G. Y. 1995. 'Fibrinogen and Cardiovascular Disorders'. *Quarterly Journal of Medicine* 88: 155–65.

11. Woeber, K. A., 1992. 'Thyrotoxicosis and the Heart'. *New England Journal of Medicine* 327: 94–8.

12. Peters, A., Ehlers, M., Blank, B., Exler, D., Falk, C., Kohlmann, T., Fruehwald-Schultes, B., Wellhoener, P., Kerner, W., and Fehm, H. L., 2000. 'Excess Triodothyronine as a Risk Factor for Coronary Events'. *Archives of Internal Medicine* 160: 1993–9.

13. Manning, J. T., and Stewart, A., 2003. 'Finger Ratios and Thyroid Function in Children from the North West Province of China'. Forthcoming.

14. Gateley, C. A., 1998. 'Male Breast Disease'. *The Breast* 7: 121–7.

15. Ferlay, J., Bray, F., Pisani, P., and Parkin, D. M., 2001. *Globocan 2000: Cancer Incidence, Mortality and Prevalence World-Wide.* Lyon: IARC Press.

16. Pisani, P., Maxwell Parkin, D., Bray, F., and Ferlay, J., 1999. 'Estimates of the Worldwide Mortality from 25 Cancers in 1990'. *International Journal of Cancer* 3: 18–29.

17. Nelson, L. R., and Bulun, S. E., 1991. 'Estrogen Production and Action'. *Journal of American Academy of Dermatology* 45 (3 suppl.): 116–24.

18. Boyd Eaton, S., Pike, M. C., Short, R. V., Lee, N. C., Hatcher, R. A., Wood, J. W., Wothman, C. M., Blurton-Jones, N. G., Konner, M. J., Hill, K. R., Bailey, R., and Hurtado, A. M., 1994. 'Women's Reproductive Cancers in Evolutionary Context'. *Quarterly Review of Biology* 69: 353–65.

19. Elledge, R. M., and Osborne, C. K., 1997. 'Oestrogen Receptors and Breast Cancer'. *BMJ* 314: 1843–4.

20. Trichopoulos, D., 1990. 'Hypothesis: Does Breast Cancer Originate in Utero?' *Lancet* 335: 1604.

21. Manning, J. T., and Leinster, S. M., 2001. '2nd to 4th Digit Ratio and Age at Presentation of Breast Cancer'. *The Breast* 10: 355–7.

Fingers and Infectious Diseases

1. Williams, H. C. (ed.), 2003. *Atopic Dermatitis: the Epidemiology, Causes and Prevention of Atopic Eczema.* New York: Cambridge University Press.

2. Chu, H. W., Honour, J. M., Rawlinson, C. A., Harbeck, R. J., Martin, R. J., 2003. 'Effects of Respiratory *Mycoplasma Pneumoniae* Infection on Allergen-Induced Bronchial Hyper-Responsiveness and Lung Inflammation'. *Infection and Immunity* 71: 1520–6.

3. Murdock, G. P., 1967. *Ethnographic Atlas.* Pittsburgh: University of Pittsburgh Press.

4. Low, B. S., 2000. *Why Sex Matters: a Darwinian Look at Human Behavior.* Princeton: Princeton University Press.

5. Unaids, 2000. Report on the global HIV/AIDS epidemic.

6. Gilks, C. F., 1999. 'The Challenge of HIV/AIDS in Sub-Saharan Africa'. *Journal of the Royal College of Physicians, London* 33: 180–4.

7. Logie, D., 1999. 'AIDS Cuts Life Expectancy in Sub-Saharan Africa by a Quarter'. *BMJ* 319: 806.

8. Bogaert, A. F., and Hershberger, S., 1999. 'The Relation between Sexual Orientation and Penile Size'. *Archives of Sexual Behavior* 28: 213–21.

9. Geschwind, N., and Galaburda, A. M., 1985. 'Cerebral Lateralisation. Biological Mechanisms, Associations and Pathology: I. A Hypothesis and a Program for Research'. *Archives of Neurology* 42: 428–59.

10. Manning, J. T., Henzi, P., Venkatramana, P., Martin, S., and Singh, D., 2003. 'Second to Fourth Digit Ratio: Ethnic Differences and Family Size in English, Indian and South African Populations'. *Annals of Human Biology* 30: 579–88.

11. Mackintosh, J. A., 2001. 'The Antimicrobial Properties of Melanocytes, Melanosomes and Melanin and the Evolution of Black Skin'. *Journal of Theoretical Biology* 211: 101–13.

12. Manning, J. T., Bundred, P. E., and Henzi, P., 2003. 'Melanin and HIV in Sub-Saharan Africa'. *Journal of Theoretical Biology* 223: 131–3.

13. De Roda Husman, A. M., and Schuitemaker, H., 1998. 'Chemokine Receptors and the Clinical Course of HIV-1 Infection'. *Trends in Microbiology* 6: 244–9.

14. Manning, J. T., Henzi, P., and Bundred, P. E., 2001. 'The ratio of 2nd to 4th Digit Length: a Proxy for Testosterone, and Susceptibility to HIV and AIDS?' *Medical Hypotheses* 57: 761–3.

15. Hooper, E., 1999. *The River: a Journey Back to the Source of HIV and AIDS.* London: Allen Lane.

16. Moore, S. L., and Wilson, K., 2002. 'Parasites as a Viability Cost of Sexual Selection in Natural Populations of Mammals'. *Science* 2015–2008.

17. Butterworth, M. B., McClellan, B., and Alansmith, M., 1967. 'Influence of Sex on Immunoglobulin Levels'. *Nature* 214: 1224–5.

18. Rowe, D. S., McGregor, I. A., Smith, S. J., Hall, P., and Williams, K., 1968. 'Plasma Immunoglobulin Concentrations in a West African (Gambian) Community and in a Healthy group of British Adults'. *Clinical and Experimental Immunology* 3: 63–79.

19. Ainbender, E., Weisinger, R., Hevitzy, M., and Hodes, H. L., 1968. 'Differences in Immunoglobulin Class of Polio Antibody in the Serum of Men and Women'. *Journal of Immunology* 101: 92–8.

20. Patty, D. W., Furesz, J., and Boucher, D. W., 1967. 'Measles Antibodies as Related to HLA Types in Multiple Sclerosis'. *Neurology* 26: 651–5.

21. Spencer, M. J., Chery, J. D., Powell, K. R., Mickey, M. R., Teraski, P. I., Mary, S. M., and Sumaya, C. V., 1977. 'Antibody Responses Following Rubella Immunization Analysed by HLA and ABO Types'. *Immunogenetics* 4: 365–72.

22. Brabin, L. and Brabin, B. J., 1992. 'Parasitic Infections in Women and Their Consequences'. *Advances in Parasitology* 31: 1–81.

23. World Health Organisation, 1987. *Expert Committee on Onchocerciasis.* World Health Organisation Technical Report Series, no. 725, Geneva.

Fingers and Skin Colour

1. Manning, J. T., Bundred, P. E., and Mather, F. M., 2004. 'Second to Fourth Digit Ratio, Sexual Selection, and Skin Color'. *Evolution and Human Behavior* 25: 38–50.

2. Sturm, R. A., Teasdale, R. D., and Box, N. F., 2001. 'Human Pigmentation Genes: Identification, Structure and Consequences of Polymorphic Variation'. *Gene* 277: 49–62.

3. Kollias, N., Malallah, Y. H., Al-Ajmi, H., Baqer, A., Johnson, B. E., and González, S., 1996. 'Erythema and Melanogenesis Action Spectra in Heavily Pigmented Individuals as Compared to Fair-Skinned Caucasians'. *Photodermatology and Photoimmunology Photomedicine* 12: 183–8.

4. Jablonski, N. G., and Chaplin, G., 2000. 'The Evolution of Skin Coloration'. *Journal of Human Evolution* 39: 57–106.

5. Holick, M. F., MacLaughlin, J. A., and Doppelt, S. H., 1981. 'Regulation of Cutaneous Previtamin D3 Photosynthesis in Man: Skin Pigment Is Not an Essential Regulator'. *Science* 211: 590–3.

6. Robins, A. H., 1991. *Biological Perspectives on Human Pigmentation.* Cambridge: Cambridge University Press.

7. Diamond, J., 1991. *The Rise and Fall of the Third Chimpanzee.* London: Random House.

8. Murdock, G. P., 1967. *Ethnographic Atlas.* Pittsburgh: University of Pittsburgh Press.

9. Mackintosh, J. A., 2001. 'The Antimicrobial Properties of Melanocytes, Melanosomes and Melanin and the Evolution of Black Skin'. *Journal of Theoretical Biology* 211: 101–13.

10. Caldwell, J., and Caldwell, P., 1996. 'The African AIDS Epidemic'. *Scientific American* 274: 62–9.

11. Szabo, R., and Short, R. V., 2000. 'How Does Male Circumcision Protect against HIV Infection?' *BMJ* 320: 1592–4.

12. Bailey, R. C., Plummer, F. A., and Moses, S., 2001. 'Male Circumcision and HIV Prevention: Current Knowledge and Future Research Directions'. *Lancet Infectious Diseases* 1(4): 223–31.

13. Weiss, H. A., Quigley, M. A., and Hayes, R. L., 2000. 'Male Circumcision, Risk of HIV Infection in Sub-Saharan Africa: a Systematic Review and Meta-Analysis'. *AIDS* 14: 2361–70.

14. Iwata, M., Corn, T., Iwata, S., Everrett, M. A., and Fuller, B. B., 1990. 'The Relationship between Tyrosinase Activity and Skin Color in Human Foreskins'. *Journal of Investigative Dermatology* 95: 9–15.

15. Manning, J. T., Bundred, P. E., and Henzi, P., 2003. 'Melanin and HIV in Sub-Saharan Africa'. *Journal of Theoretical Biology* 223: 131–3.

16. Jones, D., 1996. 'An Evolutionary Perspective on Physical Attractiveness'. *Evolutionary Anthropology* 5: 97–109.

17. Jee, S. H., Lee, S. Y., Chiu, H. C., Chang, C. C., and Chen, T. J., 1994. 'Effects of Estrogen and Estrogen Receptor in Normal Human Melanocytes'. *Biochemical Biophysical Research Communications* 199: 1407–12.

18. Wilson, N. J., and Spaziani, E., 1976. 'The Melanogenic Response to Testosterone in Scrotal Epidermis: Effects on Tyrosinase Activity and Protein Synthesis'. *Acta Endocrinology (Copenh.)* 81: 435–48.

19. Wilson, M. J., 1983. 'Inhibition of Development of Both Androgen-Dependent and Androgen-Independent Pigment Cells in Scrotal Skin Dermis of the Rat by Anti-Androgen Treatment during Foetal Growth'. *Endocrinology* 112: 321–5.

20. Van den Berghe, P. L., and Frost, P., 1986. 'Skin Color Preference, Sexual Dimorphism and Sexual Selection: a Case of Gene Culture Co-Evolution?' *Ethnic and Racial Studies* 9: 87–113.

21. Winkler, E. M., and Christiansen, K., 1993. 'Sex Hormone Levels and Body Hair Growth in !Kung San and Kavango men from Namibia'. *American Journal of Physical Anthropology* 92: 155–64.

22. Ross, R., Bernstein, L., Judd, H. R., Hanisch, R., Pike, M., and Henderson, B., 1986. 'Serum Testosterone Levels in Healthy Young Black and White Men'. *Journal of the National Cancer Institute* 76: 45–48.

23. Ellis, L., and Nyborg, H. 1992. 'Racial/Ethnic Variations in Male Testosterone Levels: a Probable Contributor to Group Differences in Health'. *Steroids* 57: 72–5.

24. Manning, J. T., 2002. *Digit Ratio: a Pointer to Fertility, Behavior and Health.* New Jersey: Rutgers University Press.

Fingers, Running Speed and Football Ability

1. Manning, J. T., and Sturt, D., 2004. '2nd to 4th Digit Ratio and Strength in Men'. Forthcoming.

2. Manning, J. T., 2002. *Digit Ratio: a Pointer to Fertility, Behavior and Health.* Rutgers University Press.

3. Manning, J. T., Bundred, P. E., and Taylor, R., 2003. 'The Ratio of 2nd to 4th Digit Length: a Prenatal Correlate of Ability in Sport'. T. Reilly and M. Marfell-Jones (eds), *Kinanthropometry VIII: Proceedings of the 12th Commonwealth International Sports Conference*, pp. 165–74. London: Routledge.

4. Manning, J. T., 2003. 'The Ratio of 2nd to 4th Digit Length and Performance in Skiing'. *The Journal of Sports Medicine and Physical Fitness* 42: 446–50.

5. Manning, J. T., and Taylor, R., 2001. 'Second to Fourth Digit Ratio and Male Ability in Sport: Implications for Sexual Selection in Humans'. *Evolution and Human Behavior* 22: 61–9.

6. Szymanski, S., 2000. 'A Market Test for Discrimination in the English Professional Soccer Leagues'. *Journal of Political Economy* 108: 590–603.

Fingers and Sexual Attraction

1. Buss, D. M., 1985. 'Human Mate Selection'. *American Scientist* 73: 47–51.

2. Gray, H., 1858. *Gray's Anatomy.* Bristol: Paragon Press.

3. Sorell, W., 1968. *The Story of the Human Hand.* London: Weidenfeld and Nicholson.

4. Schultz, A. H., 1924. 'Growth Studies on Primates Bearing upon Man's Evolution'. *American Journal of Physical Anthropology* 7: 149–64.

5. Winkler, E. M., and Christiansen, K., 1993. 'Sex Hormone Levels and Body Hair Growth in !Kung San and Kavango Men from Namibia'. *American Journal of Physical Anthropology* 92: 155–64.

6. Barkow, J. H., 1984. 'The Distance between Genes and Culture'. *Journal of Anthropological Research* 40: 367–79.

7. Belsky, J., Steinberg, L., and Draper, P., 1991. 'Childhood Experience, Interpersonal Development and Reproductive Strategy: an Evolutionary Theory of Socialization'. *Child Development* 62: 647–70.

8. Darwin, C., 1871. *The Descent of Man, and Selection in Relation to Sex*. London: Raven Press.

9. Manning, J. T., 1987. 'The Peacock's Train and the Age-Dependency Model of Female Choice'. *Journal of the World Pheasant Association* 12: 44–56.

10. Manning, J .T., 1989. 'Age-Advertisement and the Evolution of the Peacock's Train'. *Journal of Evolutionary Biology* 2: 379–84.

11. Manning, J. T., and Hartley, M. A., 1992. 'Symmetry and Ornamentation are Correlated in the Peacock's Train'. *Animal Behaviour* 42: 1020–1.

12. Miller, G. F., 2001. 'Evolution of Human Music through Sexual Selection'. N. L. Wallin, B. Merker and S. Brown (eds), *The Origins of Music*. New York: MIT Press.

13. Geschwind, N., and Galaburda, A. M., 1985. 'Cerebral Lateralisation. Biological Mechanisms, Associations, and Pathology: I. A Hypothesis and a Program for Research'. *Archives of Neurology* 42 (May): 428–59.

14. Geschwind, N., and Galaburda, A. M., 1985. 'Cerebral Lateralisation. Biological Mechanisms, Associations, and Pathology: II. A Hypothesis and a Program for Research'. *Archives of Neurology* 42 (June): 521–52.

15. Geschwind, N., and Galaburda, A. M., 1985. 'Cerebral Lateralisation. Biological Mechanisms, Associations, and Pathology: III. A Hypothesis and a Program for Research'. *Archives of Neurology* 42 (July): 634–54.

16. Sluming, V. A., and Manning, J. T., 2000. 'Second to Fourth Digit Ratio in Elite Musicians: Evidence for Musical Ability as an Honest Signal of Male Fitness'. *Evolution and Human Behavior* 21: 1–9.

17. Lack, D. 1978. 'The Significance of the Pair Bond and Sexual Selection in Birds'. T. H. Clutton-Brock and P. H. Harvey (eds), *Readings in Sociobiology*. San Francisco: W. H. Freeman.

Fingers and Homosexuality

1. Bell, A. P., Weinberg, M. S., and Hammersmith, S. K., 1981. *Sexual Preference: Its Development in Men and Women*. Bloomington: Indiana University Press.

2. Synnott, A. L., Fulkerson, W. J. and Lindsay, D. R., 1981. 'Sperm Output by Rams and Distribution amongst Ewes under Conditions of Continual Mating'. *Journal of Reproduction and Fertility* 61: 355–61.

3. Motluk, A., 2003. 'The Big Brother Effect'. *New Scientist* 29/3/03: 44–7.

4. LeVay, S., 1993. *The Sexual Brain*. Cambridge, MA: MIT Press.

5. Bagemihl, B., 1999. *Biological Exuberance*. New York: St Martins Press.

6. Bailey, M. J., and Pillar, R. C., 1991. 'A Genetic Study of Male Sexual Orientation'. *Archives of General Psychiatry* 12: 1089–96.

158 THE FINGER BOOK

7. Hall, L. S., and Love, C. T., 2003. 'Finger Length Ratios in Female Monozygotic Twins Discordant for Sexual Orientation'. *Archives of Sexual Behavior* 32: 23–8.

8. Blanchard, R., and Bogaert, A. F., 1996. 'Homosexuality in Men and Number of Older Brothers'. *American Journal of Psychiatry* 153: 27–31.

9. Blanchard, R., 1997. 'Birth Order and Sibling Sex Ratio in Homosexual and Heterosexual Males and Females'. *Annual Review of Sex Research* 8: 27–67.

10. Sanders, G., and Ross-Field, L., 1986. 'Sexual Orientation and Visuo-Spatial Ability'. *Brain and Cognition* 5: 280–90.

11. Bogaert, A. F., and Hershberger, S., 1999. 'The Relation between Sexual Orientation and Penile Size'. *Archives of Sexual Behavior* 28: 213–21.

12. Perelle, I. B., and Ehrman, L., 1994. 'An International Study of Human Handedness: the Data'. *Behavior Genetics* 24: 217–27.

13. Zucker, K. J., Beaulieu, N., Bradley, S. J., Grimshaw, G. M., and Wilcox, A., 2001. 'Handedness in Boys with Gender Identity Disorder'. *Journal of Child Psychology and Psychiatry* 42: 767–76.

14. Manning, J. T., Trivers, R. L., Thornhill, R., and Singh, D., 2000. 'The 2nd to 4th Digit Ratio and Asymmetry of Hand Performance in Jamaican Children'. *Laterality* 5: 121–32.

15. Lalumière, M. L., Blanchard, R., and Zucker, K. J., 2000. 'Sexual Orientation and Handedness in Men and Women. A Meta-Analysis'. *Psychological Bulletin* 126: 575–92.

16. Robinson, S. J., and Manning, J. T., 2000. 'The Ratio of 2nd to 4th Digit Length and Male Homosexuality'. *Evolution and Human Behavior* 21: 333–45.

17. Rahman, Q., and Wilson, G. D., 2003. 'Sexual Orientation and the 2nd to 4th Finger Length Ratio: Evidence for Organising Effects of Sex Hormones or Developmental Instability?' *Psychoneuroendocrinology*. 28: 288–303.

18. Williams, T. J., Pepitone, M. E., Christensen, S. E., Cooke, B. M., Huberman, A. D., Breedlove, N. J., Breedlove, T. J., Jordan, C. L., and Breedlove, S. M., 2000. 'Finger-Length Ratios and Sexual Orientation'. *Nature* 404: 455–6.

19. Tortorice, J., 2001. *Written on the Body: Butch/Femme Lesbian Gender Identity and Biological Correlates.* Ph.D. thesis, Rutgers University, New Jersey.

20. Tortorice, J., 2001. 'Gender Identity, Sexual Orientation, and Second-to-Fourth Digit Ratio in Females'. Paper presented at the annual meeting of the Human Behaviour and Evolution Society, London, England.

21. Brown, W., Finn, C. J., Cooke, B. M., and Breedlove, S. M., 2002. 'Differences in Finger Length Ratios between Self-Identified 'Butch' and 'Femme' Lesbians'. *Archives of Sexual Behavior* 31: 123–8.

22. McFadden, D., and Shubel, E., 2002. 'Relative Lengths of Fingers and toes in Human Males and Females'. *Hormones and Behavior* 42: 492–500.

23. Lippa, R., 2003. 'Are 2D:4D Finger-Length Ratios Related to Sexual Orientation? Yes for Men, No for Women'. *Journal of Personality and Social Psychology* 85: 179–88.

24. Manning, J. T., Churchill, A. J. G., and Peters, M., 2007. 'The Effects of Sex, Ethnicity and Sexual Orientation on Self-measured Digit Ratio (2D:4D)'. *Archives of Sexual Behaviour* 36: 222–33.

Fingers, Schizophrenia and the Feminised Ape

1. Crow, T. J., 1997. 'Is Schizophrenia the Price that Homo Sapiens Pays for Language?' *Schizophrenia Research* 28: 127–41.

2. Horrobin, D. F., 2001. *The Madness of Adam and Eve: How Schizophrenia Shaped Humanity.* London: Bantam Press.

3. Jablensky, A., 1997. 'The 100-Year Epidemiology of Schizophrenia'. *Schizophrenia Research* 28: 111–25.

4. Horrobin, D. F., 1999. 'A Speculative Overview: the Relationship between Phospholipid Spectrum Disorders and Human Evolution'. M. Peet, I. Glen and D. F. Horrobin (eds), *Phospholipid Spectrum Disorder in Psychiatry.* London: Marius Press.

5. Karlsson, J. L., 1966. *The Biologic Basis of Schizophrenia.* Springfield, IL: Thomas.

6. Seddon, B. M., and McManus, I. C., 1991. 'The Incidence of Left-Handedness: a Meta-Analysis'. Unpublished manuscript cited by Lippa, R. A., 2003. *Archives of Sexual Behavior* 32: 103–14.

7. Manning, J. T., Trivers, R. L., Thornhill, R., and Singh, D., 2000. 'The 2nd to 4th Digit Ratio and Asymmetry of Hand Performance in Jamaican Children'. *Laterality* 5: 121–32.

8. Chomsky, N., 1995. *The Minimalist Program.* Hong Kong: MIT Press.

9. Varley, R., 1995. 'Lexical Semantic Deficits following Right Hemisphere Damage: Evidence from Verbal Fluency Tasks'. *European Journal of Disorders of Communication* 30: 362–71.

10. Manning, J. T., 2002. *Digit Ratio: a Pointer to Fertility, Behavior, and Health.* New Jersey: Rutgers University Press.

11. Claridge, 1997. *Schizotypy: Implications for Illness and Health.* Oxford: Oxford University Press.

12. Stevens, J., 2002. 'Schizophrenia: Reproductive Hormones and the Brain'. *American Journal of Psychiatry* 159: 713–19.

13. Chang, C., Kokontis, J., and Liao, S., 1988. 'Molecular Cloning of Human and Rat Complementary DNA Encoding Androgen Receptors'. *Science* 240: 324–6.

14. Djian, P., Hancock, J. M., and Chana, H. S., 1996. 'Codon Repeats in Genes Associated with Human Diseases: Fewer Repeats in the Genes of Nonhuman Primates and Nucleotide Substitutions Concentrated at the Sites of Reiteration'. *Proceedings of the National Academy of Sciences of the USA* 93: 417–21.

15. Manning, J. T., Bundred, P. E., Newton, D. J., and Flanagan, B. F., 2003. 'The 2nd to 4th Digit Ratio and Variation in the Androgen Receptor Gene'. *Evolution and Human Behavior* 24: 399–405

16. Sartor, O., Zheng, Q., and Eastham, J. A., 1999. 'Androgen Receptor Gene CAG Repeat Length Varies in a Race-Specific Fashion in Men without Cancer'. *Urology* 53: 378–80.

17. Murdock, G. P., 1967. *Ethnographic Atlas*. University of Pittsburgh Press.

Index